Notes on PRACTICAL PHYSICS

MAXWELL PRESS

Notes on PRACTICAL PHYSICS

C.G. BARKLA, D.SC., F.R.S

Winner of Nobel Prize for Physics

Professor of Natural Philosophy,
University of Edinburgh

G.A. CARSE, D.SC., F.R.S.E

Lecturer on Natural Philosophy,
University of Edinburgh

MAXWELL PRESS

Chennai Trichy NewDelhi

MAXWELL PRESS

This edition has been published in india by arrangement with Carson Books, UK

ISBN 978-93-90063-73-4 Maxwell Press

No. 44, Nallathambi Street,
Triplicane, Chennai 600 005

MJP 917

Publisher's Note

The legacy of a country is in its varied cultural heritage, historical literature, developments in the field of economy and science. The top nations in the world are competing in the field of science, economy and literature. This vast legacy has to be conserved and documented so that it can be bestowed to the future generation. The knowledge of this legacy is slowly getting perished in the present generation due to lack of documentation.

Keeping this in mind, the concern with retrospective acquiring of rare books has been accented recently by the burgeoning reprint industry. Maxwell Press is gratified to retrieve the rare collections with a view to bring back those books that were landmarks in their time.

In this effort, a series of rare books would be republished under the banner, "Maxwell Press". The books in the reprint series have been carefully selected for their contemporary usefulness as well as their historical importance within the intellectual. We reconstruct the book with slight enhancements made for better presentation, without affecting the contents of the original edition.

Most of the works selected for republishing covers a huge range of subjects, from history to anthropology. We believe this reprint edition will be a service to the numerous researchers and practitioners active in this fascinating field. We allow readers to experience the wonder of peering into a scholarly work of the highest order and seminal significance.

MAXWELL PRESS

PREFACE

The following notes were written primarily for the use of the Students in the Junior Practical Physics Class in the University of Edinburgh; it is hoped they will be found useful for the Senior Classes in Schools, especially for those preparing for entrance to the University.

These notes give the general outline of the methods to be pursued. Detailed descriptions have been purposely avoided, as they frequently obscure the essential features of an experiment, while knowledge of apparatus and of its use can best be obtained in the laboratory. Sufficient detail is, however, given to enable a student with some knowledge of Physics to carry out the experiments with a minimum of attention from a teacher or demonstrator. In general, only such experiments have been described as involve the use of simple apparatus; this has, in some cases, necessitated the omission of better methods which, though simple, require apparatus of more expensive type.

The chapter on "Treatment of Observations and Determination of Error" is intended for students with more advanced knowledge of the subject, and may be omitted by those beginning the study of Practical Physics.

We are indebted to Mr W. G. Pye of Cambridge for the use of blocks of Figures 3, 4, 5, 13, 28, 45, 58.

C. G. B.
G. A. C.

University of Edinburgh,
1st January 1915.

PREFACE

CONTENTS

I.—TREATMENT OF OBSERVATIONS AND DETERMINATION OF ERROR.

II.—DESCRIPTION AND MODE OF USE OF MEASURING INSTRUMENTS.

III.—Dynamics and General Physics.

IV.—HEAT.

CONTENTS

V.—ELECTRICITY AND MAGNETISM.

VI.—LIGHT.

VII.—Sound.

VIII.—Tables.

NOTES ON PRACTICAL PHYSICS

I

TREATMENT OF OBSERVATIONS AND DETERMINATION OF ERROR.

THIS chapter is not intended to be an exhaustive treatment of the question of Errors and Accuracy, but will be found sufficient for the simple experiments in the text.

In many physical experiments we deal with a series of observations of two quantities which vary, and the object of the experiment is to test a known law, or to find the law or approximate law relating the two quantities—*e.g.*, the relation between the pressure and volume of a gas kept at constant temperature (Boyle's Law)—or we may wish to determine a constant such as a density. In each case it becomes a matter of importance to know how to treat the observations experimentally obtained, and to ascertain to what degree of accuracy the result may be supposed to be correct.

If the law is known and is expressed algebraically by a statement that some function of the quantities under discussion has a constant or approximately constant value, then by substituting the observed values we can determine this constant and see how the different values of the constant obtained from the pairs of observations agree among themselves.

If the constant thus obtained is (as it usually will be) only an approximate constant, further investigation as regards the question of errors enables us to determine whether the variations existing among the calculated values of the constant can be explained by taking account of these errors.

Certain of these errors can be avoided, but others may be due to causes beyond our control.

Errors due to known causes can be at least approximately calculated and allowed for; these are termed systematic or calculable errors, while those whose exact causes cannot be foreseen and may be unknown, are called accidental errors. Systematic errors then can be eliminated approximately. The subsequent discussion given here refers to accidental errors.

The accidental error in a determination is made up of a number of constituent errors which are small and variable in magnitude and in sign. Since the constituents may vary in sign, and thus the positive and negative parts partly balance one another, the resulting accidental error in general is also small. Hence if we make a number of determinations of the same measurement under the same conditions, and take the arithmetic mean of these, the errors tend to neutralise each other, since the frequency of occurrence of positive and negative errors of the given magnitude may be taken to be the same. The mean is thus a better approximation to the true value than a single determination; and this approximation is the closer, the greater the number of observations made. This averaging may be carried out arithmetically, or by means of a graph on which the observational points are plotted. By drawing a *smooth* curve through these points instead of an irregular zigzag curve (this irregularity being presumably due to the errors), we so far eliminate graphically the effect of these errors.

POSSIBLE ERROR OF AN OBSERVATION.

1. A Direct Observation.

To obtain the possible error of a single direct observation, it is sufficient to repeat it a number of times, keeping all the conditions the same. The mean of the results is taken, and the greatest deviation of all the values from this mean gives the possible error, provided the number of observations be sufficiently great. Thus, if the length of a tube be measured several times, and 7·40, 7·35, 7·40, 7·30, 7·45 be the successive values, we may regard the mean of these, viz., 7·38, as the true value, and the deviations from the mean as the errors.

The greatest deviation from the mean is ·08, and thus *an approximation* to the possible error or error limit is ·08. We can indicate that any observation may be out by ·08 by writing that observation as say 7·40 ± ·08. The possible error is here expressed absolutely, but it may also be expressed as a fraction of the mean—the fractional error in this case $\frac{8}{738}$ or $\frac{1}{90}$ say—or as a percentage error, in which, case it becomes 1·1 per cent.

It is not always possible to repeat an observation under exactly the same conditions. Such a case occurs in finding

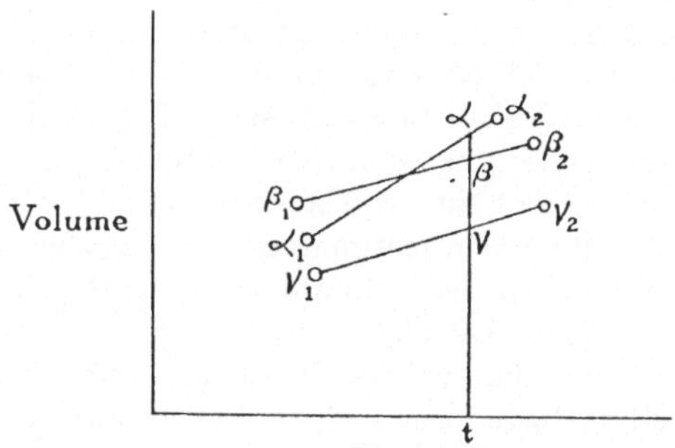

FIG. 1.

the relation between the volume and temperature of a gas kept at constant pressure. To get over the difficulty in such a case as this, several pairs of observations of temperature and volume are made, the range of temperature taken being small and approximately the same for each pair of observations. If we plot the observations, which are represented by points α_1, α_2; β_1, β_2; γ_1, γ_2, etc., and join them by lines $\alpha_1\alpha_2$, $\beta_1\beta_2$, $\gamma_1\gamma_2$, . . . then by drawing the ordinate $t\gamma\beta\alpha$ we can regard $t\alpha$, $t\beta$, $t\gamma$ as the values of the volume corresponding to the temperature t, and from these we can proceed, as already explained, to find the possible error in volume for a given temperature. Similarly, by drawing a line through some value of v parallel to the temperature axis, we can find the possible error in the temperature for a particular volume.

2. An Indirect Observation.

The possible error of an indirect observation can be obtained approximately, just as in the case of a direct observation by making a number of determinations of the quantity, *e.g.*, density. These determinations are of course calculated from direct observations of other quantities (*e.g.*, in the case of density, mass, and volume), and the maximum deviation from the mean is then found. This gives an experimentally determined possible error of a single observation (of say, density), which is an approximation to the actual error, and is in general less than the possible error computed by assuming that actual errors of observation, etc., in the components are all the greatest possible, and of such sign as to lead to the maximum error in the quantity determined. When this has been done, we get the largest error possible in an indirect observation. Clearly, in estimating the possible error of an indirect observation, we should deal with the difference between the values calculated from the *true* values of the components, and the values in which the possible errors are supposed to have occurred, in such a way as to give the maximum value to the computed error. A sufficiently close approximation will in general be got by using observed instead of true values.

EXPRESSIONS FOR POSSIBLE ERRORS OF INDIRECT OBSERVATIONS.

1. Relation of the form $y=ax$ or $y/x=a$ (*e.g.*, Hooke's Law, relation between liquid pressure and depth, pressure or volume of a gas and its absolute temperature, flexure and load, etc.).

If α = possible error in x,

β = " " y,

λ = " " a,

then
$$\lambda = \frac{y+\beta}{x-\alpha} - \frac{y}{x},$$
$$= \frac{y+\beta}{x\left(1-\frac{\alpha}{x}\right)} - \frac{y}{x} = \frac{\beta}{x} + \frac{y}{x^2}\alpha,$$

neglecting quantities of the second order (*i.e.*, a^2 and $a\beta$)

$$= \frac{y}{x}\left(\frac{a}{x} + \frac{\beta}{y}\right);$$

hence $\frac{\lambda}{a} = \frac{a}{x} + \frac{\beta}{y}$ = fractional error = sum of fractional errors in x and y.

2. Relation of form $xy = a$ (*e.g.*, Boyle's Law).

In this case $\lambda = (x + a)\ (y + \beta) - xy$

$$= \beta x + ay = xy\left(\frac{\beta}{y} + \frac{a}{x}\right);$$

hence $\frac{\lambda}{a} = \frac{\beta}{y} + \frac{a}{x}$ = fractional error = sum of fractional errors in x and y.

3. Relation of the form $y^2/x = a$ (*e.g.*, relation between time and length of simple pendulum).

$$\lambda = \frac{(y+\beta)^2}{x-a} - \frac{y^2}{x} = \frac{(y+\beta)^2}{x\left(1 - \frac{a}{x}\right)} - \frac{y^2}{x} = 2\beta\frac{y}{x} + \frac{y^2}{x^2}a$$

$$= \frac{y^2}{x}\left(\frac{a}{x} + \frac{2\beta}{y}\right)$$

also $\frac{\lambda}{a} = \frac{a}{x} + \frac{2\beta}{y}$.

Note that since the square of y occurs in the expression, the effect on the result of an error in y is twice that of an error in x.

If $a = x^p y^q$, p and q having any values, positive or negative, integral or fractional,

$$\frac{\lambda}{a} = p\frac{a}{x} + q\frac{\beta}{y},$$

the signs of a and β being chosen so as to make the result a maximum.

The results can also be worked out by differentiation, thus :—

If the relation between the observed quantities $x, y, z, \ldots$ is $f(x, y, z, \ldots) = a$ where a is the constant whose value is to be determined, then we can apply the processes of the differential calculus to determine the error in a resulting from the various errors in $x, y, z, \ldots$ We can consider the possible errors in $x, y, z, \ldots$ as small increments of $x, y, z, \ldots$ and denote them by $\Delta x, \Delta y, \Delta z, \ldots$

$$\begin{aligned} \text{Then} \quad a+\Delta a &= f(x+\Delta x, y+\Delta y, z+\Delta z, \ldots) \\ &= f(x, y, z, \ldots)+\Delta x f_x+\Delta y f_y+\Delta z f_z+\ldots; \\ \therefore \quad \Delta a &= f_x \Delta x+f_y \Delta y+f_z \Delta z+\ldots \end{aligned}$$

neglecting small quantities of the second order.

Since the signs of $\Delta x, \Delta y, \Delta z \ldots$ may be either positive or negative, in the final result the signs must be so adjusted that Δa may have a maximum value.

$$\left.\begin{aligned} E.g., \quad a &= xy \\ \text{then} \quad \Delta a &= x\Delta y + y\Delta x \\ \text{and} \quad \frac{\Delta a}{a} &= \frac{\Delta x}{x} + \frac{\Delta y}{y} \end{aligned}\right\} \text{as in No. 2.}$$

POSSIBLE ERROR OF A MEAN IN TERMS OF THAT OF A SINGLE OBSERVATION.

We can compute the possible error of a mean by the above rules.

$$\text{If } m=\frac{m_1+m_2+\ldots m_n}{n}$$

$$\lambda = \text{P.E. of } m$$

$$\alpha = \text{P.E. of } m_1, m_2 \ldots m_n,$$

$$\text{then } \lambda=\frac{(m_1+\alpha)+(m_2+\alpha)+\ldots+(m_n+\alpha)}{n}-\frac{m_1+m_2+\ldots m_n}{n}$$

$$=\alpha$$

on the assumption that all the α's have the same sign and all the maximum magnitude. This, however, is highly improbable, and it can be shown from the theory of probabilities that the probable error of a mean of n observations is the probable error of a single observation divided by the square root of n.

Doubtful Observations.—It may happen that if a constant is calculated from, or a graph drawn of two varying quantities, a value or point is very divergent from the others. In such a case the observations round about that particular value should be repeated, and if the new observations fall in line with the others as regards the value of the constant, or the points on the graph, then the student may fairly conclude that the abnormal observation may be neglected. Observations which seem abnormal should, however, in general only be rejected after careful consideration.

INVESTIGATION OF LAWS.

The investigation or verification of laws may be carried out by (1) an arithmetical method, (2) a graphical one, (3) a combination of the arithmetical and graphical methods.

1. Arithmetical Method.

By tabulating the values of the two quantities x and y, between which a relation is to be found, it is often possible to obtain the relation by mere inspection. Thus, if we take the following series of corresponding values of x and y,

	y	$\frac{y}{x}$	Deviation from Mean (value of $\frac{y}{x}$ - mean).
1	·788	·788	+·005
2	1·544	·772	−·011
3	2·327	·776	−·007
4	3·169	·792	+·009
5	3·902	·780	−·003
6	4·744	·791	+·008
7	5·568	·795	+·012
8	6·213	·777	−·006
9	6·985	·776	−·007
10	7·841	·784	+·001
		Mean = ·783	

we see at once that x and y increase together, and also that when x is doubled, trebled, etc., then y is also doubled, trebled, etc. Thus the relation y/x = constant is suggested; hence the various values of y/x are tabulated in the third column, and the deviation from the mean in the fourth column. If y/x were only approximately constant, its value would gradually change from one end of the column to the other; but if the deviations from the mean be due to accidental errors, they in general, as here, occur both positively and negatively, and form an irregular variation. In the last column the signs of the deviations are distributed irregularly, while positives and negatives occur with equal frequency; moreover, there is no evidence of a *gradual* change in the magnitude of the deviation.

In other and more complex cases it may be observed that while $y=0$ when $x=0$, y/x decreases as x increases, i.e., y^n/x may be constant (where $n > 1$). This suggests that some higher power of y, say y^2, y^3, etc., should be tried, and the law y^2/x = constant, or y^3/x = constant, etc., should be tested.

2. Graphical Method.

The above process may become laborious if a complex function of the quantities x and y is involved, and then recourse may be made to the *graphical method.* It may be possible by plotting a graph of the quantities x and y to recognise the form of it. If the graph is a straight line, then its equation will be of the form $y=ax$, or $y=ax+b$, according as it does or does not pass through the origin. Similarly, a parabola of the form $y^2/x=a$, or a rectangular hyperbola of the form $xy=a$, when the asymptotes are the axes, may be recognised and tested, while the student who has studied graphs will with comparative ease recognise other more complex graphs. In the case in which $y^n=ax$, then since taking logarithms, $n \log y=\log a+\log x$ we see that $\log y$ plotted against $\log x$ will give a straight line, and from its slope the value of n can be obtained; while the value of $\log a$ and therefore a can be obtained from an intercept. In general, the above graphs are sufficient for any of the laws to be found or tested in the experiments given.

3. Combination of Arithmetical and Graphical Methods.

The graph may suggest the law, and then the law can be quickly tested by the arithmetical method.

CONSISTENCY OF OBSERVATIONS WITH LAWS.

We have still to test the consistency of the observations with the law found or suggested. We have to show that the calculated values of the function do not deviate from the mean value by more than can be accounted for by the experimental errors, and also that these deviations do not show any *gradual* change in the value of the function from one end of the series to the other.

We shall deal with the case in which the tests are applied arithmetically, though it is to be noted that they may also be applied graphically. Take the case of the following corresponding observations of x and y, which represent the relation between the pressure and volume of a gas at constant temperature (Boyle's Law).

x = Pressure (cms. of Hg).	$y \propto$ Volume (cms. of tube).	xy.	Deviation from Mean.	Computed P.E. of xy.
45·2	20·9	945	+5	7
47·7	19·7	940	±0	7
59·0	15·8	932	−8	7
59·4*	15·8*	939	−1	7·5
59·5*	15·9*	946	+6	7·5
59·6*	15·7*	936	−4	7·5
62·9	15·0	944	+4	8
76·0	12·4	942	+2	9
94·3	9·9	934	−6	10
99·2	9·5	942	+2	11
P.E. in x =0·1 cm.	P.E. in y =0·1 cm.	Mean=940.		

The possible errors of pressure and volume were got from the three marked observations in columns 1 and 2 respectively, and the possible error of the function xy was computed from these as in example 2, p. 5. We see that there is no indication in the second last column of any gradual change in the value of the function, and the distribution of signs is a random one. Thus, the second of the two tests for consistency is satisfied, and by considering the last two columns we see that all the divergencies (except one) are covered by the possible errors, and hence we may fairly conclude that the first test is also satisfied and that the law found is xy = constant.

II

DESCRIPTION AND MODE OF USE OF MEASURING INSTRUMENTS.

In physical experiments the measurement of a length is of frequent occurrence, and it is essential to know what instruments are available and suitable for carrying out the required measurement. There are several instruments which can be used, the simplest being a graduated scale, while the majority of others are more complicated on account of having refinements and adjustments by which the measurement of length can be made with greater accuracy.

The Measuring Rod consists of a lath of boxwood one metre or half a metre in length, subdivided into centimetres and millimetres, or of a steel rule of a suitable length with millimetre or half-millimetre divisions. If the scale can be applied directly to the body to be measured, the readings may be estimated by the eye to a fraction of a millimetre, care being taken to avoid errors due to parallax. If the points between which a length is measured are not in contact with the scale, the scale should be placed parallel to the line joining them, and should be viewed in a direction perpendicular to its length.

The Beam Compass consists of a rod along which two slides with points are movable, the slides being capable of being clamped by means of two screws. The points are adjusted by means of the slides till they coincide with the two points the distance between which is required, and the distance can then be read off the beam compass directly if it is graduated, or indirectly by measuring the distance between the points of the slides by means of a measuring scale. In practice the slides are fixed so that the distance between them is very nearly equal to that of the distance to be measured, the mark of one of the slides also being made to coincide with a division of the beam compass scale, and with one end of the length to be measured, then by means of a fine screw the other slide is adjusted so that its point coincides with the other end of the length to be measured. This slide usually has a vernier attached to it, and thus the measurement of length can be made with comparative accuracy.

The Vernier is a useful adjunct which enables greater accuracy to be obtained, and obviates the necessity of employing a scale with inconveniently small divisions. It consists of a slide, capable of moving along the scale, and graduated so that say $n \pm 1$ of its divisions are equal to n of those of the scale. If we consider the common case, in which ten vernier divisions are equal to nine scale divisions, then since $10\,v = 9\,s$, therefore $s - v = s - \frac{9}{10}\,s = \frac{1}{10}\,s$, thus the difference between a scale division and a vernier division is $\frac{1}{10}$ of a scale division. If, then, the zero of the vernier coincides with one of the ends of the length to be measured, and lies between two of the divisions of the scale, then we see that some vernier division will approximately coincide with a scale division; and if, as in the figure, the second vernier division coincides with a scale

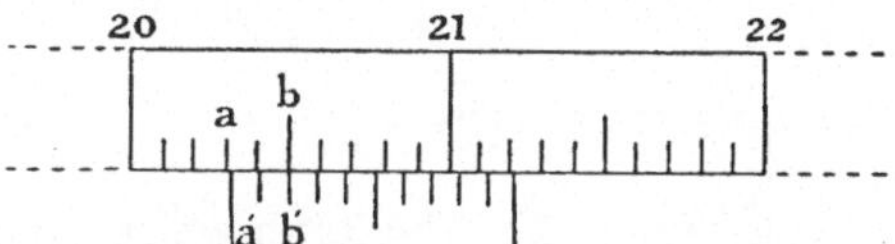

FIG. 2.

division, then the distance aa' which we wish to measure is $ab - a'b' = 2s - 2v = \cdot 2$ of the small scale divisions, *i.e.*, millimetres, and the reading is 20·32 cms.; hence by observing the number of the vernier division which coincides with a scale division, we can read the scale to a tenth of one of its actual divisions. Other forms of verniers will be found on various instruments, *e.g.*, microscopes, sextants, etc., the readings of which the student should be able to work out for himself.

Calipers.—When a measuring scale cannot be applied directly to a length to be measured, then calipers may be used. These consist of two varieties, inside and outside, which are used for inside and outside measurements respectively. Calipers consist of two arms hinged at one end; inside calipers have straight arms with points bent outwards, and the outside ones have curved arms with points bent inwards. The measurements are made by adjusting the points of the calipers to the required distance, and then reading the distance between their points by means of a scale.

Vernier Calipers consist of a graduated rod furnished with two jaws at right angles to the rod—one fixed at the end, the

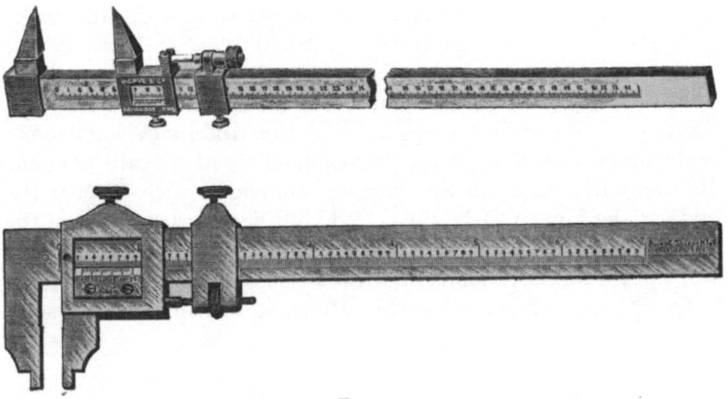

FIG. 3.

other movable along the rod. The latter which carries a vernier is capable of being clamped in any position; a fine adjustment may be made by means of a screw on the slide.

The Screw Gauge is useful for measuring small lengths,

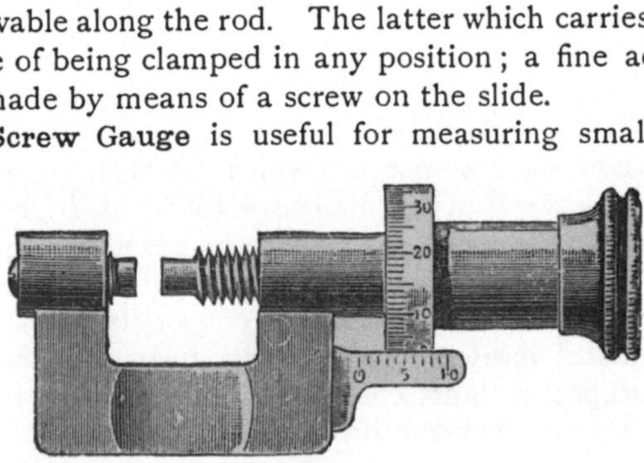

FIG. 4.

(*e.g.*, the diameters of wires), and consists of a well-cut screw which works in a piece of metal with its ends turned at right angles to its length. One whole turn of the screw moves the screw-head through a distance equal to the pitch of the screw, while fractions of this distance can be read off a scale on the circumference of the screw-head.

The Spherometer consists of an equilateral tripod with a screw of known pitch, working vertically at the centroid of the triangle; the lower end of the screw carries a pointed end, and to the head of the screw is attached a circular disc with a graduated scale enabling fractions of a turn of the screw to be

FIG. 5.

read, while a small vertical scale attached to the tripod gives the vertical distance between the point of the screw and the plane containing the feet of the tripod, in terms of the pitch of the screw.

Note.—In the majority of these instruments there exists a zero error, *e.g.*, in the spherometer, when the four points are in one plane, the zero of the vertical scale and that of the graduated disc may not coincide, and the reading then got is the zero error which will have to be added or subtracted from the final reading when a measurement is made. So in the case of the vernier calipers, screw gauge, etc.

THE DIAMETER OF A WIRE OR THICKNESS OF A PLATE BY MEANS OF THE SCREW GAUGE.

Screw the jaws of the gauge up till they *just* touch, and note the zero error, if any, and its sign, *i.e.*, whether it has to be added or subtracted finally. Do this a number of times, and take the mean as the zero error. (It can be shown, see p. 6, that the probable error of a mean of n observations is $1/\sqrt{n}$ of the probable error of a single observation.)

Then insert the wire or plate whose diameter or thickness is required, and screw the jaws up till they just grip the wire or plate, and take the reading of the gauge. Do this several times at the one place, and hence obtain the possible error of a reading. Also make measurements at different positions of the wire or plate; the average of these when account is taken of the zero error will give the diameter of the wire or thickness of the plate, assumed to be uniform.

THE THICKNESS OF A PLATE BY THE SPHEROMETER.

An accurately plane surface, usually of a glass plate, is used for adjusting the pointed screw, so that the four points (the three points of the legs of the tripod and the point of the screw) are in the same plane. This is done by screwing the central leg till the instrument is just not capable of rotation on the glass surface about the central leg. The zero error is then read and its sign noted. The plate whose thickness is required is then placed under the central leg, which has been screwed up to allow the plate to slide underneath it. The central leg is then screwed down till it just touches the plate, the rotation test being applied, and the thickness is read off from the vertical scale and disc, account being taken of the zero correction. As in the case of the screw gauge, the possible error of a measurement and the mean thickness of the plate can be determined.

THE RADIUS OF CURVATURE OF A SPHERICAL SURFACE BY THE SPHEROMETER.

The zero error of the instrument is observed, and the instrument is placed on the surface of a lens, say, and the central leg adjusted by the rotation method till it is just in contact with the lens.

Let the reading corrected for the zero error given by the spherometer be p, and let R be the radius of curvature of the lens surface. Then as shown in the figure, $p(2R-p)=r^2$, where r is the radius of the circle passing through the three feet of the spherometer.

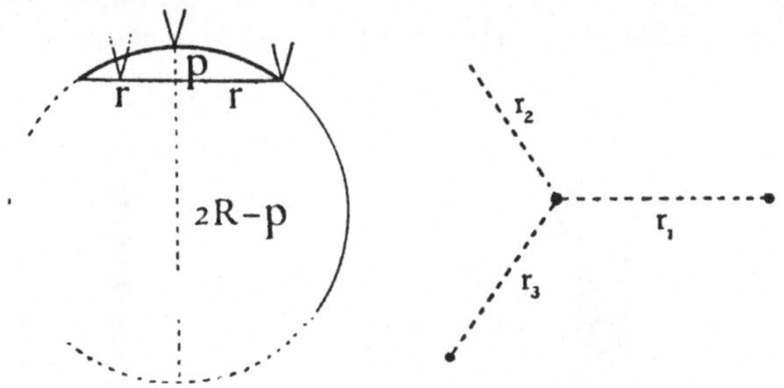

FIG. 6.

To get r, adjust the central point to the same plane as the three fixed points, and press the spherometer on a sheet of paper, making impressions of the points. Measure by vernier calipers the distance from the central point to each of the outer points, and take the mean value of r.

Substituting for p and r in the above equation we get R.

FORTIN'S BAROMETER.

This consists of a glass tube 80 cm. or more in length closed at the upper end, and dipping into a cistern of mercury at the lower end (Fig. 7). The tube being free of air, is filled to a height of seventy odd centimetres with mercury. Above this is vapour of mercury which is in a very rarefied state and produces a negligible pressure. The mercury in the tube is supported by the pressure of the atmosphere on the surface of the mercury in the cistern, this pressure being communicated through the mercury to the mercury in the tube at the same level. The height of the column of mercury above the level in the cistern is a measure of the atmospheric pressure.

Before reading the height, the level in the cistern is adjusted to the zero of a fixed measuring scale by raising or lowering the bottom of the cistern which is of flexible leather. The adjustment is made by a screw until the surface of mercury just touches a fixed ivory pointer. (This is best done by observing the pointer and its image in the mercury surface.) The scale by the side of the tube gives distance from the end of the pointer—that is, after adjustment from the level of the surface. Near the top of the tube is a vernier scale which moves in front of the barometer tube and alongside the principal scale, and which is capable of adjustment by means of a screw at the side.

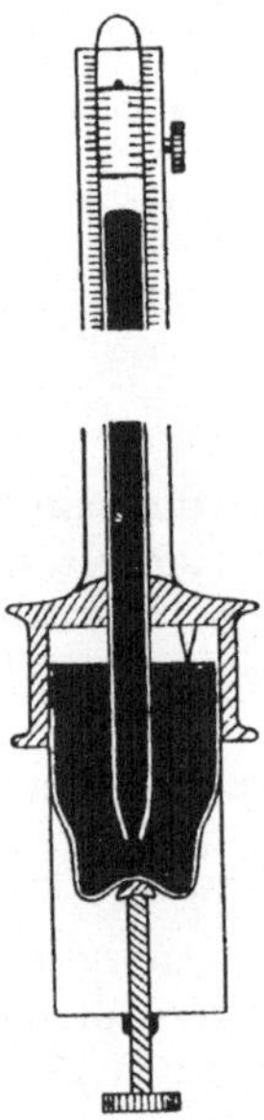

FIG. 7.

FIG. 8.

The zero of the scale is at its lower edge, and a second edge on the other side of the barometer tube moves with the vernier scale in front, so that the two lower edges give the horizontal plane through the zero of the vernier scale. In adjusting the scale, therefore, the eye should be placed in a position so that the two edges are superposed, and in the plane of the mercury surface. The vernier reading should then be taken.

The variations in the height of the barometer, due to temperature changes and consequent changes in density of the mercury and in length of the scale, are usually negligible. It should, however, be remembered that the height of a barometer depends upon its altitude above sea-level, so that if barometers are to be compared their readings should be corrected to readings at the same level—say sea-level.

SIPHON BAROMETER.

In the siphon barometer the cistern is formed by the lower portion of the glass tube turned up parallel to itself, as in Fig. 8. The end of the short limb is also closed, but the air pressure is communicated through a small aperture in the side. The difference between the levels in the two limbs is the height of the barometer. This is determined by means of two scales, which have a common zero somewhere near the middle of the longer limb, one reading upward and the other downward. The sum of the readings of the two surfaces thus gives the distance between them. The scales may be engraved on the tube, or on glass, or metal beside the tube, and sliding vernier scales are sometimes attached for greater accuracy of reading.

THE BALANCE.

In determining mass among the factors that may have to be taken into account are:—

(*a*) Errors in the weights,
(*b*) Inequality of the lengths of the arms of the balance,
(*c*) Buoyancy of the air.

(*a*) It is assumed that the weights are correct, or that the errors produced in the case of a good set of weights are negligible in ordinary determinations.

(*b*) The error due to the arms of the balance being of unequal length may be eliminated by (1) substitution, or (2) double weighing.

In the first method the body to be weighed is placed in one pan of the balance, and counterpoised by shot or sand or weights in the other pan. The body is then removed, the counterpoising masses being left on the other pan, and weights are substituted for the body till there is balance. The weights thus substituted for the body give its weight.

In the case of double weighing, the body is weighed first in the one pan and then in the other, and the true value of the weight is the square root of the product of its weight in the two pans. If the arms of the balance are not very unequal, it will in general be sufficient to take the arithmetic mean of the apparent weights.

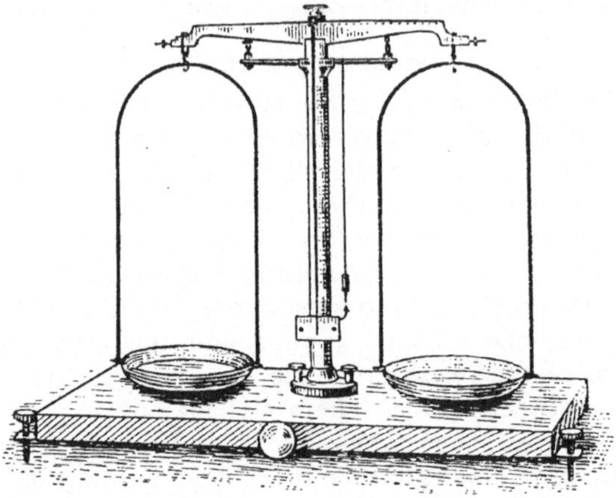

FIG. 9.

It is clear that the ratio of the lengths of the arms of the balance may also be obtained from the observations in double weighing, for if l_1 and l_2 be the lengths of the arms and w_1 and w_2 the weights *balancing* the body suspended from arms of length l_2 and l_1 respectively, and W the true weight of the body,

then
$$w_1 l_1 = W l_2$$
$$W l_1 = w_2 l_2,$$

whence
$$W^2 = w_1 w_2,$$

and
$$\frac{l_1}{l_2} = \left(\frac{w_2}{w_1}\right)^{\frac{1}{2}}.$$

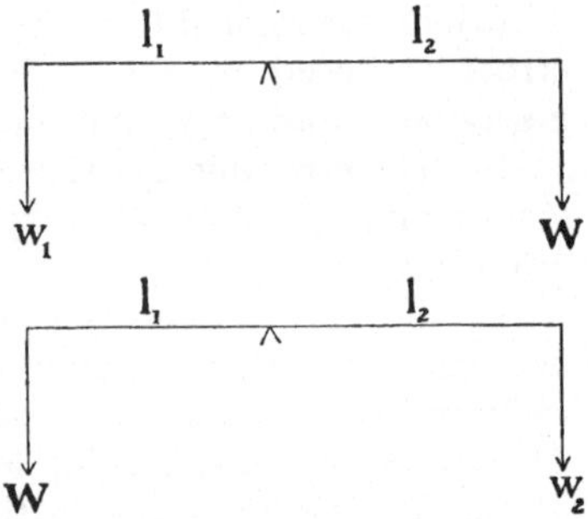

FIG. 10.

(*c*) If as is in general the case, the body which is being weighed has not the same density as the weights, then since

the weighing is being made in air, different quantities of air will be displaced by the body and weights respectively, hence the true weight of the body, *i.e.*, its weight if the weighing had been made in a vacuum, is not directly ascertained. A correction can, however, be made which reduces the weighing to weighing *in vacuo.* For if W and w be the true weights *in vacuo* of the body and weights which balance it in air, and Δ and δ the densities of the body and of the material of the weights respectively, and σ the density of the air under the conditions of the experiment, then for equilibrium
weight of body − buoyancy of air on body
= weight of weights − buoyancy of air on weights,

$$i.e., \quad W - \sigma\frac{W}{\Delta} = w - \sigma\frac{w}{\delta},$$

whence approximately $W = w(1 - \frac{\sigma}{\delta} + \frac{\sigma}{\Delta}).$

Thus the correction per gram of weight is $\sigma(1/\Delta - 1/\delta)$, and is positive or negative according as Δ is $<$ or $>\delta$.

In the following table are tabulated values of $\sigma(1/\Delta - 1/\delta)$ for brass weights ($\delta = 8{\cdot}4$) for various values of Δ.

Buoyancy Correction for Weighings, with Brass Weights, in Air of Density ·0012 g. per cubic centimetre.

Density of Body Weighed.	Correction, in mg. per g. of Weight.	Density of Body Weighed.	Correction, in mg. per g. of Weight.
g. per c.c.		g. per c.c.	
0·25	+4·66	2·5 (glass)	+0·34
0·5	+2·26	3·0	+0·26
0·6	+1·86	4·0	+0·16
0·7	+1·57	5·0	+0·10
0·8	+1·36	6·0	+0·06
0·9	+1·19	7·0 (iron)	+0·03
1·0 (water)	+1·06	8·0	+0·01
1·1	+0·95	8·4 (brass)	0·00
1·2	+0·86	9·0	−0·01
1·3	+0·78	10·0	−0·02
1·4	+0·71	12·0	−0·04
1·5	+0·66	13·6 (mercury)	−0·05
1·6	+0·61	14·0	−0·06
1·7	+0·56	16·0	−0·07
1·8	+0·52	18·0	−0·08
1·9	+0·49	20·0	−0·08
2·0	+0·46	22·0	−0·09

III

DYNAMICS AND GENERAL PHYSICS.

DENSITY OF A SOLID BY DETERMINATION OF MASS AND

(a) Direct Determination of Volume.

Apparatus.—Solid whose density is required (metal in the form of cube, cylinder, or other regular shape), calipers, measuring rod, balance, weights.

Method.—Find the mass M of the solid by weighing, and the volume V by calculation from its dimensions, then the quotient M/V gives the density—in the case of the C.G.S. system of units, as so many grams per cubic centimetre.

(b) Indirect Determination of Volume.

Apparatus.—Solid whose density is required (not necessarily of regular shape), measuring glass, balance, weights.

Method.—Find the mass M of the solid by weighing, and the volume V by immersing it in a quantity of water which is sufficient to cover it in the measuring glass, and noting the reading on the measuring glass of the surface of the water before and after the body is introduced. The difference of the two readings gives the volume. The density is then found as above.

SPECIFIC GRAVITY OF A SOLID BY THE PRINCIPLE OF ARCHIMEDES.

(a) Solid which sinks in Water.

Apparatus.—Solid whose specific gravity is required, balance, weights, beaker.

Method.—Weigh the solid in air (W_a). Weigh it when immersed in the beaker of water (W_w); this is done by suspending the body from the end of the balance beam by means of a fine thread or wire (whose weight can be taken into account if necessary), so that it is wholly immersed in the water, and finding its weight. The beaker is supported by means of a stool so placed in the balance case as to be clear of the pan.

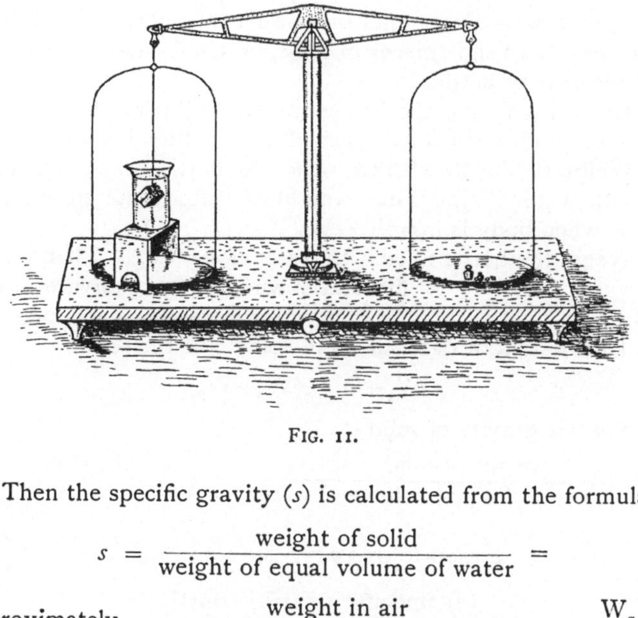

FIG. 11.

Then the specific gravity (s) is calculated from the formula

$$s = \frac{\text{weight of solid}}{\text{weight of equal volume of water}} =$$

$$\text{approximately} \quad \frac{\text{weight in air}}{\text{weight in air—weight in water}} = \frac{W_a}{W_a - W_w}$$

[Buoyancy of the air is neglected.]

(b) Solid that floats in Water.

Apparatus.—As above; sinker.

Method.—Weigh the body in air (W_a); weigh body and sinker in water (W_w); weigh sinker in water W'_w.

Then weight of body alone in water $= W_w - W'_w$, which is a negative quantity. Thus the upward thrust of water on the body is $W_a - (W_w - W'_w)$, which is the weight of an equal volume of water.

Specific gravity of body

$$= \frac{\text{weight of body}}{\text{weight of equal volume of water}} = \frac{W_a}{W_a - W_w + W'_w}.$$

SPECIFIC GRAVITY OF SUBSTANCE IN SMALL FRAGMENTS OR A POWDER BY THE SPECIFIC GRAVITY BOTTLE.

Apparatus.—Fragments or powder of substance insoluble in water—(lead shot, pieces of glass, or sand)—balance, weights, specific gravity bottle.

Method.—Weigh the bottle empty (W_1); weigh the bottle with the solid in it (W_2); thus weight of solid $= W_2 - W_1$.

Weigh the bottle with the solid, the bottle now being filled up with water (W_3). Thus, weight of water required to fill the bottle when body is in $= W_3 - W_2$.

Weigh the bottle filled with water alone (W_4). Then weight of water required to fill bottle when it is otherwise empty $= W_4 - W_1$.

Thus weight of water of same volume as the solid

$$= W_4 - W_1 - (W_3 - W_2).$$

Specific gravity of solid

$$= \frac{\text{weight of solid}}{\text{weight of equal volume of water}} = \frac{W_2 - W_1}{W_4 - W_1 - W_3 + W_2}.$$

SPECIFIC GRAVITY OF A LIQUID.

(*a*) Specific Gravity Bottle.

Apparatus.—Liquid, specific gravity bottle, balance, weights.

Method.—Weigh the bottle empty (W_1); weigh the bottle filled with the liquid (W_2); weigh the bottle filled with water (W_3).

Specific gravity of the liquid

$$= \frac{\text{weight of any volume of liquid}}{\text{weight of equal volume of water}} = \frac{W_2 - W_1}{W_3 - W_1}.$$

(*b*) Hare's Apparatus.

Apparatus.—Liquid, water, Hare's apparatus.

Method.—Adjust the tubes so that their ends dip into the vessels containing the liquid and water respectively. By means of the rubber tube draw air out so that the liquid rises in each tube, and then clamp the rubber tubing. Measure the respective heights (h_1, h_2) of the liquid and water columns above the levels of the surfaces of the liquids in the vessels; then if ρ_1

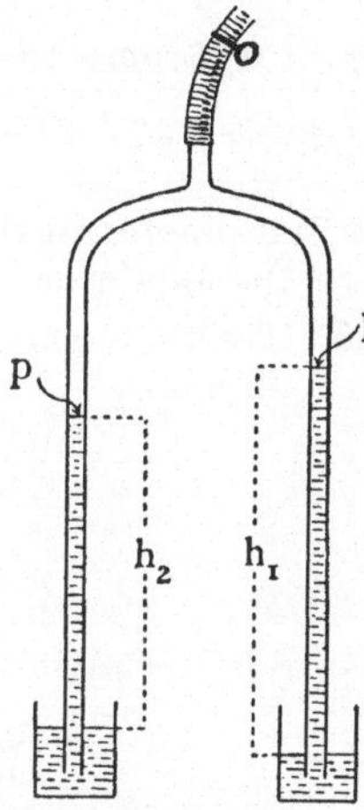

FIG. 12.

and ρ_2 be the densities of the liquid and water respectively, the atmospheric pressure $\Pi = p + g\rho_1 h_1$, where p is the pressure of the air above the liquid columns. This also $= p + g\rho_2 h_2$;

$$\therefore \quad g\rho_1 h_1 = g\rho_2 h_2.$$

$$\frac{\rho_1}{\rho_2} = \frac{h_2}{h_1}; \quad \therefore \quad \frac{s_1}{s_2} = \frac{h_2}{h_1}$$

where s_1 and s_2 are the specific gravities.

Taking s_2 as 1 for water, $\qquad s_1 = \dfrac{h_2}{h_1}.$

(c) Upthrust Method.

Apparatus.—Liquid, a body on which the liquid has no chemical action (say a glass bulb), balance, weights.

Method.—Weigh the body in air (W_1); weigh the body in the liquid (W_2); weigh the body in water (W_3).

Thus, weight of liquid equal in volume to that of body $= W_1 - W_2$.

" " water " " " $= W_1 - W_3$.

Specific gravity of liquid

$$= \frac{\text{weight of volume of liquid}}{\text{weight of equal volume of water}} = \frac{W_1 - W_2}{W_1 - W_3}.$$

(*d*) **Common Hydrometer** (Fig. 13).

Apparatus.—Liquid, glass vessel, hydrometer (with suitable range of values).

Method.—Float the hydrometer in the liquid and take the reading of the surface of the liquid on the hydrometer, which directly gives the specific gravity.

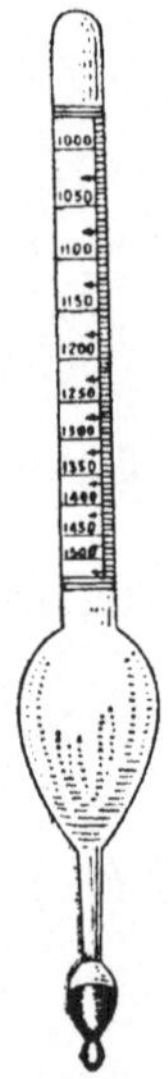

FIG. 13.

FIG. 14.

(*e*) **Nicholson's Hydrometer** (Fig. 14).

Apparatus.—Liquid, deep glass vessel, Nicholson's hydrometer, balance, weights.

Method.—Find the weight of the hydrometer (W_1). Float the hydrometer in the liquid, and by adding weights (W_2) to the upper pan, adjust till the standard mark comes to the surface of the liquid. Thus the total weight of liquid displaced by the hydrometer $= W_1 + W_2$.

Do the same with the hydrometer in water, the weight added being W_3. Then total weight of water displaced by the hydrometer $= W_1 + W_3$.

Specific gravity of liquid

$$= \frac{\text{weight of any volume of liquid}}{\text{weight of equal volume of water}} = \frac{W_1 + W_2}{W_1 + W_3}.$$

RELATION BETWEEN DISTANCE AND TIME IN THE CASE OF MOTION ON AN INCLINED PLANE.

Apparatus.—Smooth board, metal cylinder or ball, small wooden block, stop-watch.

Method.—Put the wooden block under one end of the board, so that the latter is slightly inclined. From a mark at the higher end of the plane, allow the metal cylinder or ball to roll from rest down the plane, and measure the times (t) that it takes to roll various distances (s) down the plane.

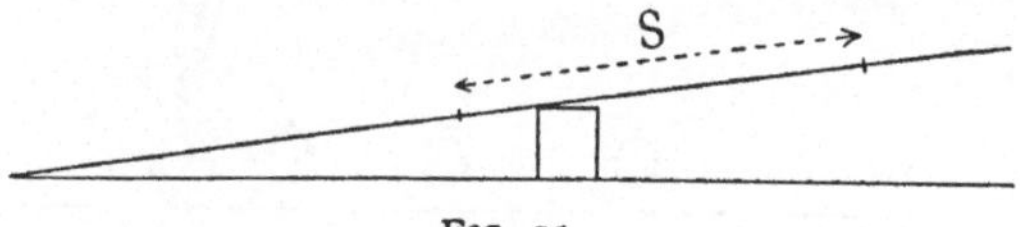

FIG. 15.

Tabulate the values of s and t, and both graphically and by calculation show that s/t^2 is constant for a given inclination.

The experiment may be repeated with the board at different inclinations, and the accelerations (a) determined from the relation for uniformly accelerated motion

$$s = \frac{1}{2}at^2; \qquad \frac{s}{t^2} = \frac{1}{2}a.$$

Show from these determinations that a is proportional to the sine of the inclination of the plane to the horizontal.

PARALLELOGRAM AND TRIANGLE OF FORCES
(Equilibrium of Three Forces).

Apparatus. — Drawing board and paper, three spring balances, cords.

Method.—To the hook of each balance a cord is attached, and the other ends of these cords are fastened to a ring O. Two of the balances are fastened to two hooks A and B

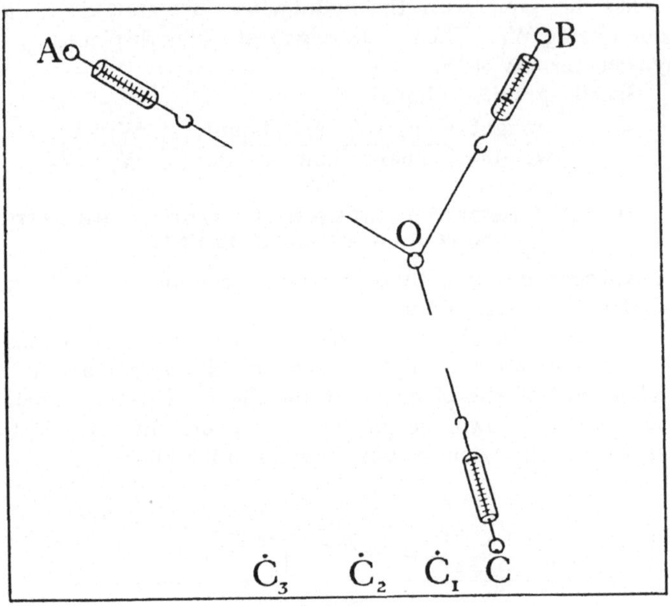

FIG. 16.

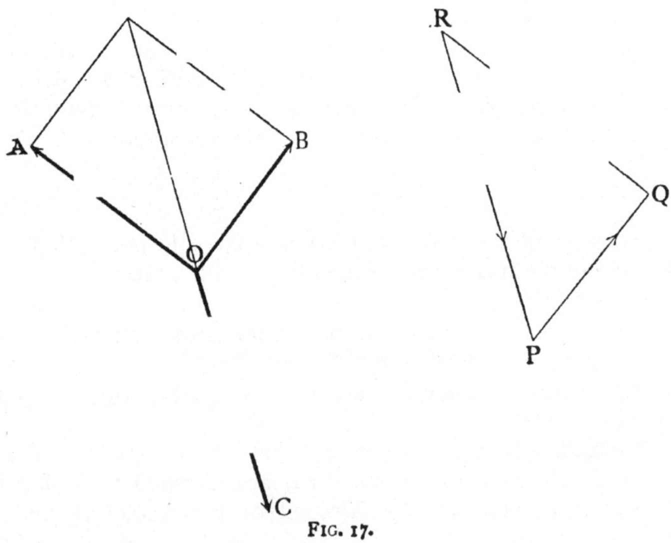

FIG. 17.

in the corners of the drawing board, while the end C of the third balance is moved along the lower edge of the drawing board into positions C, C_1, C_2, etc. (Fig. 16).

Mark on the paper the directions OA, OB, OC, and construct a diagram in which OA, OB, and OC represent in magnitude and direction the forces exerted by the spring balances, and show that a triangle PQR can be constructed with sides equal and parallel to OB, OA, and OC.

Do this for the various positions of C.

Vary the method by graphically finding the resultant of two of the forces. Show that this is equal in magnitude but opposite in direction to the third force (Fig. 17).

INCLINED PLANE (Mechanical Advantage with Force Acting (1) Up the Plane, (2) Horizontally).

Apparatus.—Inclined plane, trolley or roller with scale-pan attached by string passing over pulley, weights.

1. Force up the Plane.

Method.—The trolley A is loaded with a certain weight, and weights are added to scale-pan B till the trolley just begins to move up the plane. Take readings of the total

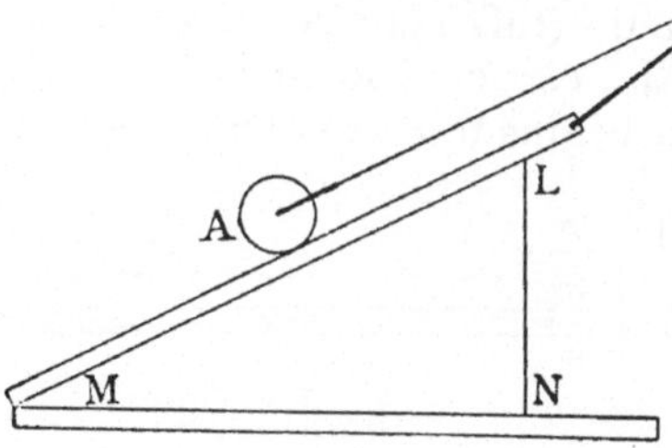

FIG. 18.

weights of A and B and of the distances LN and LM. Find also the weight of B when the trolley just begins to move down the plane, the inclination of the plane being the same. Vary the inclination of the plane and proceed as before.

Tabulate the mean of the two weights of B, say P and that of A, say W and also LN and LM. Calculate also and tabulate W/P and LM/LN. These ratios will be found to be equal; they give the mechanical advantage.

2. Force Horizontal.

Method.—Arrange the apparatus so that the force is now acting horizontally instead of parallel to the length of plane,

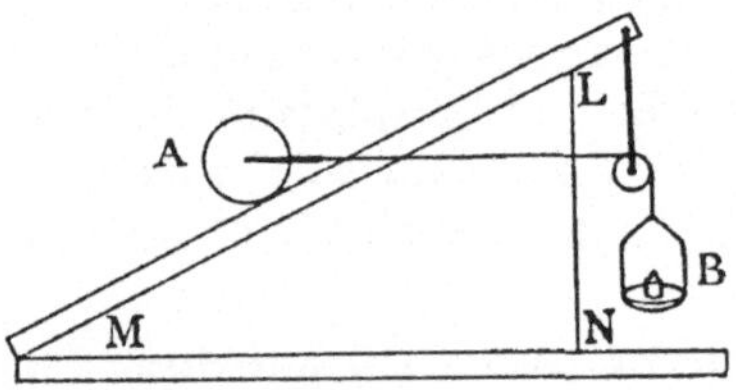

FIG. 19.

and proceed as in (1) tabulating corresponding values of P and W and LN and MN, W/P, and MN/LN.

It will be seen that W/P = MN/LN.

FRICTION.

Apparatus.—Friction board with pulley, slider, weights.

Method.—After having placed the friction board so that its surface is horizontal—(this may be tested by a spirit-level)—place weights on the carrier C so that with a given weight on the slider A it just begins to move along the friction board.

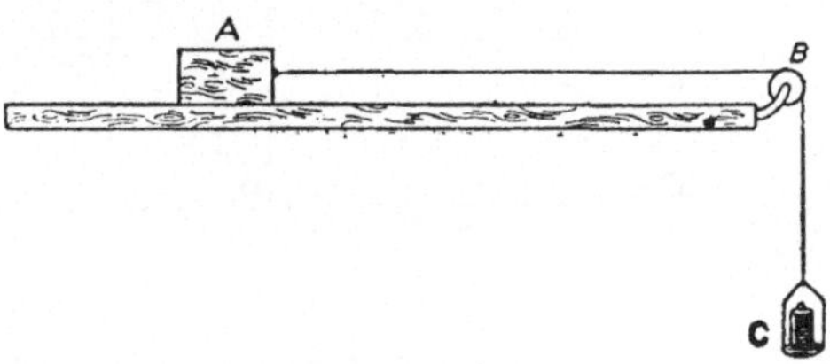

FIG. 20.

Tabulate the results, giving weight of slider + weights, and force required to move the slider from rest.

Repeat the experiment, but instead of finding the force required to start the slider from rest, find what force is necessary to keep the slider moving *with uniform speed* after it has been started.

It will be found that the force which is required to start the slider from rest is greater than that required to maintain a uniform velocity when it has been started. Static friction is thus shown to be greater than kinetic friction.

To Investigate the Variation of Friction with Normal Pressure. (Coefficients of Friction.)

Proceed as in the above investigation and tabulate the values of friction (F) and normal pressure (R) in each case.

The graphs of these are straight lines through the origin. This can further be tested by calculating the coefficients of friction $\mu = F/R$, and showing that each is a constant for the two surfaces under consideration.

By using the other side of the slider so that the area now in contact is only half that of the first side, and by repeating the preceding experiment, it can be shown that the coefficients of friction are independent of the area of the surfaces in contact. It is, of course, essential that the second surface is as nearly as possible in the same state as the first.

Angle of Friction.

Apparatus.—Friction board with hinged base, slider, weights, wood blocks, measuring scale.

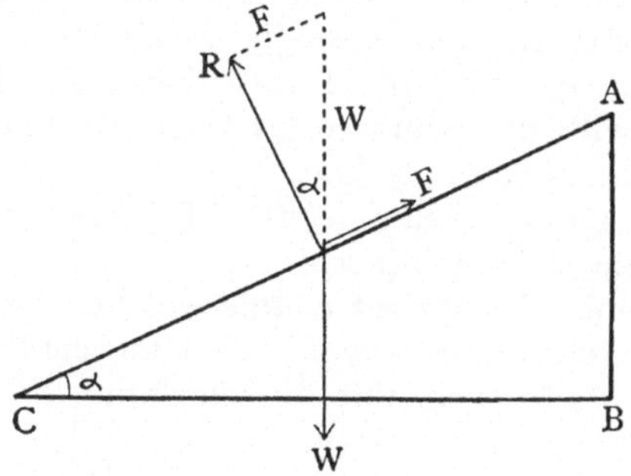

FIG. 21.

Method.—Place a weight on the slider and tilt the friction board till the slider when started moves down the plane with

uniform speed as in the experiments on kinetic friction. Find the angle by measuring the dimensions of the inclined plane, *e.g.*, perpendicular and base, and calculate the tangent of the angle. Repeat with different weights on the slider.

By the triangle of forces

$$F : R : W :: AB : BC : AC$$

$$\therefore \quad \mu = \frac{F}{R} = \frac{AB}{BC} = \tan \alpha,$$

i.e., the tangent of the angle at which sliding occurs is equal to the coefficient of friction—in this case kinetic.

If the angle α be that at which the slider *just begins* to move down the plane, the corresponding coefficient is that of static friction.

RELATION OF TIME OF SMALL OSCILLATION OF SIMPLE PENDULUM TO ITS LENGTH AND DETERMINATION OF "g."

Apparatus.—Fine wire or string, metal ball, clamp, stop-watch, measuring scale, calipers.

Method.—Attach the wire to the ball and pass its other end through the clamp, and fasten the clamp so that the wire is tightly held at the lower edge of the clamp.

Measure the length of the wire of the pendulum in centimetres, and add the radius of the bob; this is the so-called length of the pendulum (l).

Observe the time in seconds of a given number of swings (the greater the number the more precise will be the determination of the time of a single oscillation). The swings should be through an arc of less than 10°. Repeat these observations with pendulums of at least six different lengths, varying between 30 and 100 cms.

Construct a table giving length of pendulum (l cm.) and time of a single oscillation (t secs.).

Plot a graph of l against t; this will be of the form of a parabola with the origin as a point on the curve. Hence the equation is $l = at^2$. Test this by calculating l/t^2 and plotting l against t^2.

To calculate "g."—The relation just found gives $l/t^2 = a$, and it can be shown that

$$t = 2\pi\sqrt{\frac{l}{g}}$$

where t = time of a complete oscillation, *i.e.*, of a motion of the pendulum to and fro. Hence

$$g = \frac{4\pi^2 l}{t^2} = 4\pi^2 a.$$

LAW OF INCREASE OF LIQUID PRESSURE WITH DEPTH.

Apparatus.—Cylindrical glass vessel with water, graduated metal tube closed at one end, weights or pieces of metal of equal weight.

Method.—Float the metal cylinder in the water, and take the reading of the scale on the cylinder at the surface of the water. Take a series of similar readings with various weights inside the metal cylinder.

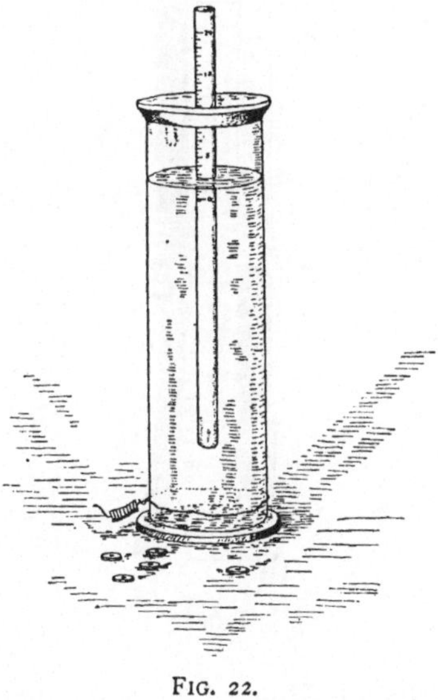

FIG. 22.

Tabulate results showing total weight (W) of cylinder and weights and depth (d) to which the cylinder sinks below the surface of the water.

Show by calculation and a graph that W/d is a constant.

(For question of accuracy and calculation of error, see Chapter I.)

BOYLE'S LAW.

Apparatus.—Boyle's law apparatus, in which pressures both greater and less than atmospheric can be used.

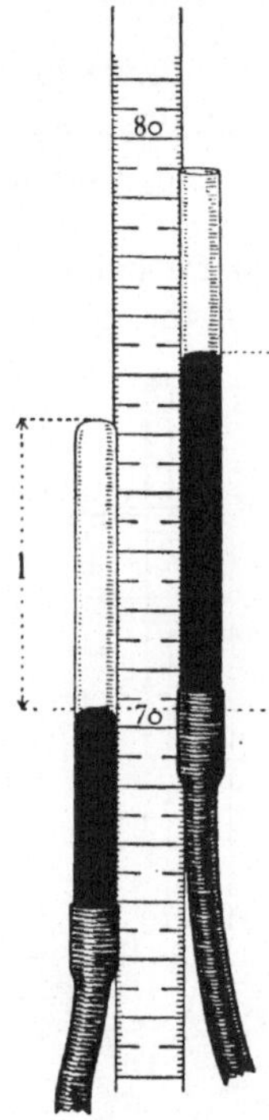

FIG. 23.

Method.—Take a reading of the barometer at the beginning and at the end of the experiment, and use the mean of these two readings in the calculation of the resultant pressure. The temperature should also be noted, as the conditions of the experiment require that the temperature should be constant. (The temperature will, in general, vary only slightly during the time of the experiment.) Observations of the simultaneous readings of the two mercury levels, and the reading of the top of the enclosed air column are made, and from these, h and l can be calculated, where h is the difference of levels of the mercury in the two arms, and l is the length of the enclosed air column. If the diameter of the closed tube is uniform, then l is proportional to the volume of the enclosed air, and the pressure is given by atmospheric pressure (H) expressed in height of mercury column $\pm h$, according as the level of the mercury in the enclosed air column is lower or higher than that in the open tube.

By raising or lowering one of the arms of the U-tube, different pressures and consequently different volumes will be obtained.

Construct a table giving the different values of the total pressures P, viz., $H \pm h$, the corresponding volumes (V), and the product of pressure and volume PV. This product will be found to be approximately constant. The graph of P and V will be a rectangular hyperbola, PV=constant, and the graph of P and 1/V will be a straight line passing through the origin. (For a discussion of the question of accuracy of the experiment, see the chapter on Accuracy, p. 1.)

SURFACE TENSION.

Relation between the Height to which Water rises in a Capillary Tube and the Bore of the Tube; and Determination of the Surface Tension.

Apparatus.—Capillary tubes of different radii, glass vessel with water, glass scale, vessel containing mercury, vernier calipers, balance, weights.

Method—Calibration of the Tubes.—The radii of the capillary tubes may be given, or it may be necessary to determine the radii. In the latter case a small thread of mercury is inserted in the tube by dipping it in a vessel of mercury, and then the upper end having been closed by the finger, tilting it into a horizontal position, when it will be found that a small quantity of mercury has remained in the tube. The mercury may be made to travel along the tube by slightly tilting, and by means of a pair of vernier calipers its length at different portions of the tube can be determined. The constancy or the variation of the lengths at different portions of the tube gives an indication of the uniformity or variation of the bore of the tube. In any case in which the lengths differ considerably among themselves, the tube should be rejected and another procured. The weight of the mercury in the tube is then ascertained, either by weighing the mercury in a watch-glass, or by weighing the tube containing the mercury, and subtracting from that weight the weight of the tube alone.

The length (l) of the mercury thread being known (the mean of the observed lengths being taken), and also its mass (m), we can compute the value of the radius (r) of the tube from the equation—

$$\text{volume} \times \text{density} = \text{mass},$$
$$i.e., \ \pi r^2 l\rho = m,$$

where ρ = density of mercury and may be taken as 13·6, and the tube is assumed of circular section. The radii of the various tubes are thus determined.

Measurement of Height to which Water rises.—Fasten the tube to the glass scale, as in the figure, by means of rubber bands, and take the reading of the top of the water in the tube

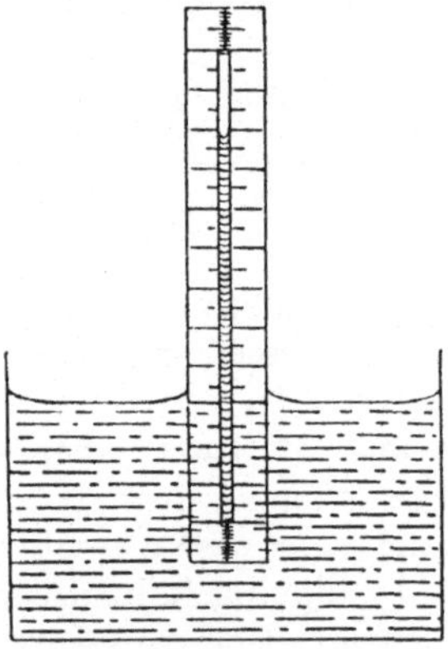

FIG. 24.

on the scale, and also the reading of the level surface by looking with the eye just below the level of the surface. The difference of these two readings gives h, the height to which the water has risen in the capillary tube. Do this for the different tubes.

Care should be taken that the water thoroughly wets the interior of the tube, say by moving the tube up and down in the water until the water rises and falls freely in the tube, when the tube is lowered and raised.

Plot a graph of h against r which will be of the form of a rectangular hyperbola, and show that rh is a constant. (For the question of the accuracy of the result, see p. 5, No. 2.)

The surface tension (T) may be calculated from the expression

$$T = \frac{rh\rho g}{2},$$

where ρ is the density of the water. In the above expression for T the volume of liquid in the meniscus has been neglected. More accurately

$$T = \frac{g\rho r}{2}(h+\tfrac{1}{3}r).$$

For a liquid, the angle of contact of which is α, the expression is

$$T = \frac{rh\rho g}{2\cos\alpha}, \text{ neglecting the volume of meniscus.}$$

It is not essential that the tube be of uniform bore, provided the radius at the meniscus can be measured. If a microscope with micrometer eye-piece is available, this may be done by filing and breaking the tube there and measuring the diameter by direct observation through the microscope.

RELATION BETWEEN ELONGATION AND LOAD FOR A SPIRAL SPRING.

Apparatus.—Spring with pointer attached near one end, weights and arrangement for suspending weights from spring, cathetometer.

Method.—Support the spring at one end with the pointer at the lower free end. Take the reading of the position of the pointer (from any arbitrary position, say the table top) when the load is zero, and for a series of different loads in increasing order of magnitude, making an occasional determination of the zero position.

When the greatest load has been put on, make a series of observations as above, with different loads in decreasing order of magnitude. Tabulate load (W), scale reading (s), total extension of spring (l), and l/W. The mean of the elongations for the same load should be used for the value of l. It will be seen that l/W is a constant, that is, the elongation is proportional to the load. (This is of course equally true if the zero position be taken when the spring supports a load. Additional elongations and additional loads take the place of elongation and load simply.)

Plot a graph of l against W. This is a straight line through the origin with equation $l=k$ W, where k is a constant.

RELATION BETWEEN ELONGATION AND LOAD FOR A RUBBER CORD.

Apparatus.—Rubber cord fastened at its upper end with pan for weights at its lower end, weights, cathetometer.

Method.—Load the pan sufficiently to straighten the cord. Take a series of readings of the distance between two marked points P and Q for different weights, (1) when the load is increasing, (2) when the load is decreasing. Allow a minute or two to elapse between putting on the load and taking the reading (Fig. 25).

The graph of elongation (l) and load (W) will be a slightly curved line, and in general the graph for decreasing load will differ slightly from that for increasing load. If the load applied be too great, the rubber cord acquires a "permanent set."

TEST OF HOOKE'S LAW FOR ELONGATION OF A WIRE, AND DETERMINATION OF YOUNG'S MODULUS.

Apparatus.—Two wires several metres long suspended side by side about a centimetre or two apart, scale and vernier, screw gauge.

Method.—One wire A has a weight hung at its lower end to keep it straight, and weights are hung from the other wire B. The elongation in B produced by the different weights is measured by the scale and vernier, the scale being fastened to A and the vernier to B. Put the maximum load which is supplied in the scale-pan of B, and leave the wire stretching for a few minutes (Fig. 26).

Make a series of observations of the position of the vernier (s) and load (W), both when the loads are applied in increasing and decreasing order of magnitude, the load being increased or decreased by equal amounts each time. The two sets of readings should agree within the limits of error, unless the greatest weight applied has been sufficient to stretch the wire beyond the "limits of elasticity," in which case the wire will have acquired a "permanent set."

FIG. 25.

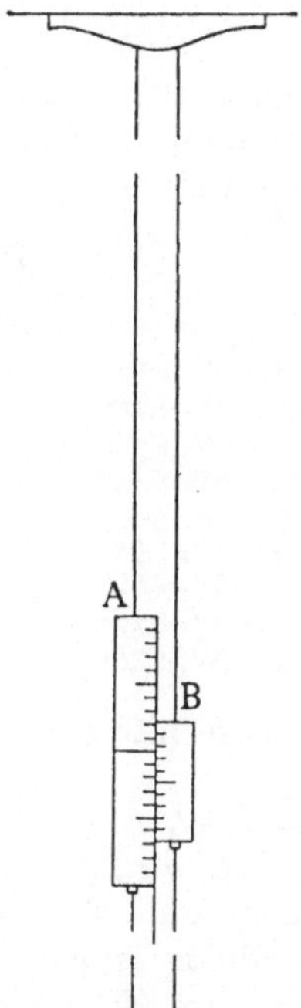

FIG. 26.

The original length of the wire from the point of support to the point where the vernier is clamped being L, and the alteration of its length caused by a load w being l, then within the limits of elasticity l/w is constant for that wire.

This can be shown from the observations either graphically or by calculation.

Young's modulus (Y) is the ratio of longitudinal stress to the corresponding fractional elongation—the strain. The stress is the force per unit area, *i.e.*, $w/\pi r^2$, where r = radius of wire; the strain is the elongation per unit length, *i.e.*, l/L.

$$\therefore \quad Y = \frac{\frac{w}{\pi r^2}}{\frac{l}{L}} = \frac{wL}{\pi r^2 l}.$$

To calculate the mean elongation l for a given load w from an even number of readings of elongation and load, say 8, tabulate them thus:—

Elongation.

	Load.	Load increasing.	Load decreasing.	Mean.
1.	W_1+1w	l'_1	l''_1	l_1
2.	W_1+2w	l'_2	l''_2	l_2
3.	W_1+3w	l'_3	l''_3	l_3
4.	W_1+4w	l'_4	l''_4	l_4
5.	W_1+5w	l'_5	l''_5	l_5
6.	W_1+6w	l'_6	l''_6	l_6
7.	W_1+7w	l'_7	l''_7	l_7
8.	W_1+8w	l'_8	l''_8	l_8

Divide these into two equal portions, the first four and the last four.

Calculate l_5-l_1, l_6-l_2, etc., elongations for a load of $(W_1+5w)-(W_1+w)$, etc., viz., $4w$, and find the mean (l) of l_5-l_1, etc., for the load of $4w$. Thus the value of l/w in the formula for Young's modulus is got, and by measuring in addition the length of the wire (L), and the radius (r), the value of Y can be calculated.

BENDING OF BARS—SUPPORTED AT ONE OR BOTH ENDS.

(a) Relation of Flexure to Load for a Given Bar.
(b) Test of the Relation of Flexure to Length for a Constant Load.

Apparatus.—Bar supported at one end by a clamp, or at both ends by wedges, carrier and weights capable of being suspended from the unclamped end or middle of the bar in the two cases, vertical scale to measure the flexure, measuring rod.

Method.—Find the zero position of the end or middle of the bar in the respective cases, and make a series of observations of the deflection (f) produced (**a**) when the weights (W) alone are varied (the weight of the carrier being taken into account) and the length kept constant, (**b**) when the length (l) is varied and the weights (W) are kept constant.

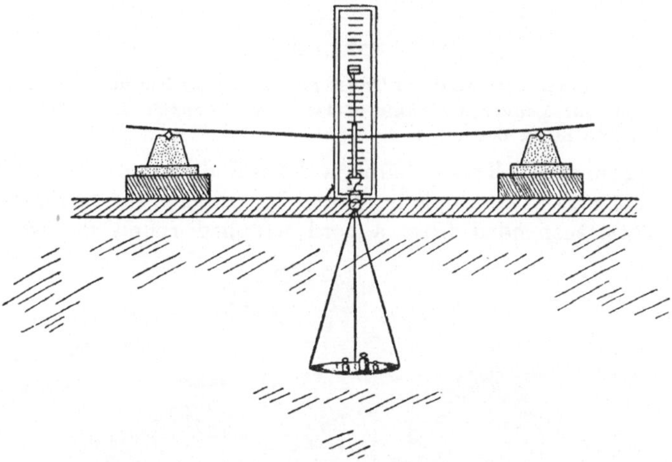

FIG. 27.

In the case of the observations (**a**) show that the ratio of deflection to weights appended is constant within fairly wide limits.

In case (**b**) the ratio of l^3/f is constant.

YOUNG'S MODULUS BY THE BENDING OF A BAR.

For a bar of *rectangular* section supported at both ends and bent by a load at its centre, Young's modulus Y is given by

$$Y = \frac{Wl^3}{4bd^3f},$$

where l, b, d are the length, breadth, and depth respectively, and f is the deflection at the centre for an increase in the load of W.

It will be noticed that if W is constant, then as shown experimentally l^3/f is constant, so that Young's modulus can be calculated, using the mean value of l^3/f obtained in that experiment.

In the case of a similar bar clamped horizontally at one end and supporting a load at the free end, the formula is

$$Y = \frac{4Wl^3}{bd^3f}.$$

TORSION.

Relation between (1) Angle of Twist and Twisting Couple for a Rod of Constant Length, (2) Angle of Twist and Length for a Rod, the Couple being Constant.

Apparatus.—Torsion lathe, weights and carrier.

Method.—The couple is applied by means of the carrier and weights, suspended from a cord wrapped round the wheel.

FIG. 28.

One end of the rod to be tested is clamped to the block, and the other end, which is free, carries the wheel. The angle of twist is measured by a pointer and scale, the pointer being attached to the rod and the scale to the base. Pointer and scale are, however, capable of adjustment anywhere along the rod.

(1) The pointer and scale are first placed at the end of the rod near the wheel and kept there, and readings are taken of the scale position of the pointer for a series of weights from zero upward. Tabulate the values of W and θ, W being the total weight including the carrier and θ the angle turned through by the pointer, and show graphically and by calculation that W/θ is constant.

(2) The weight suspended from the cord is kept constant, while a series of readings is taken with the pointer and scale in various positions along the rod, that is, for various values of l. For each value of l the scale reading is taken first when the rod is in its normal unstrained state, and second when the fixed weight is suspended from the wheel. Thus the twist θ for various values of l is obtained. Plot l and θ, and tabulate, giving values of θ/l. This will be found to be constant.

DETERMINATION OF RIGIDITY BY A STATICAL METHOD.

Apparatus.—Torsion lathe, and accessories.

Method.—It may be shown that the rigidity (n) of the material of the rod used is given by

$$n = \frac{2gpR}{\pi a^4} \times \frac{Ml}{y}$$

where p = length of the pointer measured from the axis of the rod,

y = distance along the arc through which the tip of the pointer moves when a mass M is placed in the pan,

R = radius of the wheel,

a = radius of the rod,

l = distance of the pointer from the fixed end of the rod.

The quantity $2gpR/\pi a^4$ is a constant for a given rod in the apparatus, so that it can be determined by measuring p, R, and a, and then the mean of a series of Ml/y taken. If the instrument indicates the angle (ϕ) in radians directly, the formula is

$$n = \frac{2gR}{\pi a^4} \times \frac{Ml}{\phi}.$$

RELATION OF TIME OF OSCILLATION OF TORSIONAL PENDULUM TO ITS LENGTH.

Apparatus.—Ring or cylinder with clamp at centre, wire, stop-watch, measuring scale.

Method.—Attach one end of the wire to the bob by means of the clamp, and fasten the other end of the wire to another clamp so that the pendulum can oscillate round the vertical wire as axis.

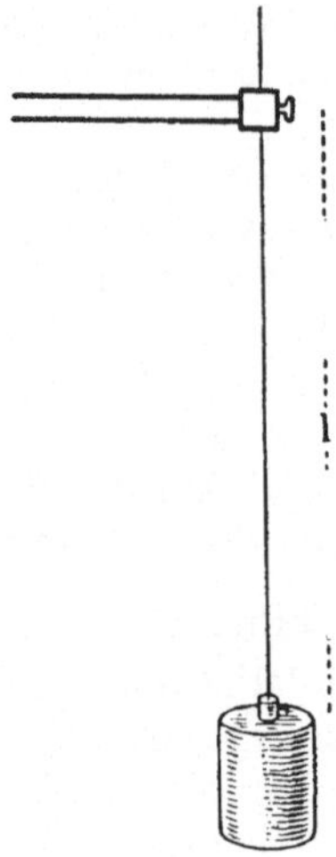

FIG. 29.

Measure the length (l) of the supporting wire in centimetres from clamp to clamp, and observe the time in seconds of a given number of swings.

Repeat these observations with wires of at least six different lengths, varying say between 30 and 100 cms.

Construct a table giving length of pendulum (l cm.) and time of a single oscillation (t secs.). Plot a graph of l against t; this will be of the form of a parabola with the origin as a point on the curve. The equation to the curve is $l=at^2$. This can be shown by calculating l/t^2, which will be found to be approximately constant. Plot l against t^2.

MEASUREMENT OF MOMENTS OF INERTIA BY THE METHOD OF OSCILLATION.

Apparatus.—Torsional pendulum, body of regular form or of known moment of inertia, stop-watch.

Method.—To find the moment of inertia (I_1) about an axis through the centre of gravity of the body forming the bob of the torsional pendulum, (1) find the time of oscillation (t_1) of the body when suspended as a torsional pendulum with a given length of wire; (2) find the time of oscillation (t_2) when the moment of inertia of the system is altered, by adding to the bob of the pendulum a body whose moment of inertia (I_2) is known or can be calculated. (The suspension is not changed, and hence the couple tending to restore the bodies to their position of equilibrium is unchanged.)

In (1) $$t_1 = 2\pi\sqrt{\frac{I_1}{c}}, \quad . \quad . \quad . \quad I_1$$

where t_1 = time of a complete oscillation, I_1 = moment of inertia of bob about the axis of vibration, c = a constant depending upon the length, diameter, and nature of the suspending wire. (It is the value of the couple which would twist one end of the wire through a radian if the other end were fixed.)

In (2) $$t_2 = 2\pi\sqrt{\frac{I_1 + I_2}{c}}$$

where I_2 = moment of inertia of the regular body about the axis of rotation whence,

$$\frac{I_1 + I_2}{I_1} = \frac{t_2^2}{t_1^2};$$

$$\therefore \quad I_1 = I_2 \frac{t_1^2}{t_2^2 - t_1^2}.$$

If the irregular body of moment of inertia (I_1) is *replaced* by the one of known moment of inertia (I_2), the same wire being used and remaining unaltered in length, then

$$(1) \quad t_1 = 2\pi\sqrt{\frac{I_1}{c}}$$

$$(2) \quad t_2 = 2\pi\sqrt{\frac{I_2}{c}};$$

$$\therefore \quad I_1 = I_2 \frac{t_1^2}{t_2^2}.$$

Hence I_1 can be determined.

RIGIDITY OF A SOLID BY TORSION (Dynamical Method).

Apparatus.—Torsional pendulum, stop-watch, measuring scale, screw gauge.

Method.—For a torsional pendulum, see p. 43.

$$t = 2\pi\sqrt{\frac{I}{c}}.$$

It can be shown from the theory of elasticity that

$$c = \frac{\pi r^4 n}{2l}$$

where l is the length of wire and r its radius, and n is the rigidity of the substance of the wire;

$$\therefore \quad n = \frac{8\pi I l}{t^2 r^4}.$$

Using a body of known moment of inertia (I) as the bob of the pendulum (the suspending wire being that of the material whose rigidity is to be measured), find the time of a complete oscillation, measure the length of the wire and its radius, and hence calculate the value of the simple rigidity.

Care should be taken to measure the radius of the wire very exactly, as the fourth power of radius of the wire occurs in the expression for the rigidity (see Chapter I.).

It is also advisable to measure the radius of the wire at several points along its length, and to take the mean of the values obtained.

IV

HEAT.

DETERMINATION OF THE BOILING POINT OF A PURE LIQUID.

Apparatus.—Flask fitted with cork having two holes in it, into one of which a thermometer can be fitted and into the other a glass tube to lead off the vapour.

Method.—Put some of the liquid whose boiling point is to be found into the flask with some glass beads or small pieces of glass to prevent "bumping." Heat the flask till the liquid boils, and read the temperature at which this occurs. The temperature observed should be that of the vapour and not that of the liquid, as the former is a constant depending only on the particular liquid and the atmospheric pressure. The latter depends slightly upon other conditions. Consequently, the thermometer should be in the vapour above the liquid and not in the liquid itself. Read on the barometer the atmospheric pressure at which the boiling point is determined.

TO DETERMINE THE BOILING POINT OF A SOLUTION.

Apparatus.—Flask fitted with cork having two holes in it, into one of which a thermometer can be fitted and into the other a glass tube to lead off the vapour.

Method. — Put say 100 grams of water into the flask and find the boiling point of the water (as in the previous experiment). Then add 10 grams of common salt and find the boiling point of the solution by having the thermometer bulb *immersed in the solution.* Find also the temperature of the vapour coming off from the salt solution. Repeat with solutions which contain 20 and 30 grams respectively, and estimate the elevation of the boiling point due to the 10,

20, etc., grams of salt added. To allow for evaporation, small quantities of water should be occasionally added. The amount may be determined by weighing roughly.

MELTING POINT OF PARAFFIN WAX (First Method).

Apparatus.—Piece of glass tube drawn into the form of a thin-walled capillary tube, thermometer, beaker.

Method.—Fill the capillary tube with small pieces of solid paraffin, and heat gently in a flame till the wax melts in the tube. Attach the tube to the stem of the thermometer by means of an elastic band, and in such a position that the wax is very near to the bulb of the thermometer. Immerse the thermometer in a beaker of water, and heat the water and continually stir it till the wax *just* becomes clear. Note the temperature at which this occurs, and allow the water gradually to cool while still being stirred, and again note the temperature at which the wax just becomes solid. The mean of these two temperatures gives the melting point of the paraffin wax.

It is advisable, in order to obtain an accurate result, to take the mean of the observations in say three experiments performed as above described.

MELTING POINT OF PARAFFIN WAX (Second Method).

Apparatus.—Paraffin wax, two beakers, thermometer.

Method.—Melt the wax in one of the beakers, and when the temperature is a little above the melting point of the wax insert the thermometer. Keep the thermometer in long enough to allow the wax which solidifies on the bulb to liquefy again. Take the thermometer out and note the temperature at which a whitish film of wax appears on the thermometer bulb. Now place the thermometer in a beaker of water, heat slowly, and note the temperature at which the film disappears.

Repeat several times (say three), and take the mean of the results. Compare the melting point got by this method with that given by the other methods described.

MELTING POINT OF A SOLID BY THE COOLING METHOD (Third Method).

Apparatus.—Paraffin wax, beaker, thermometer, stop-watch.

Method.—Heat the paraffin wax in the beaker to a temperature above that of its melting point. Insert the thermometer in the beaker, and having removed the flame, take a cooling curve by reading the temperature every minute or half-minute. Plot a graph of temperature and time. During the process of

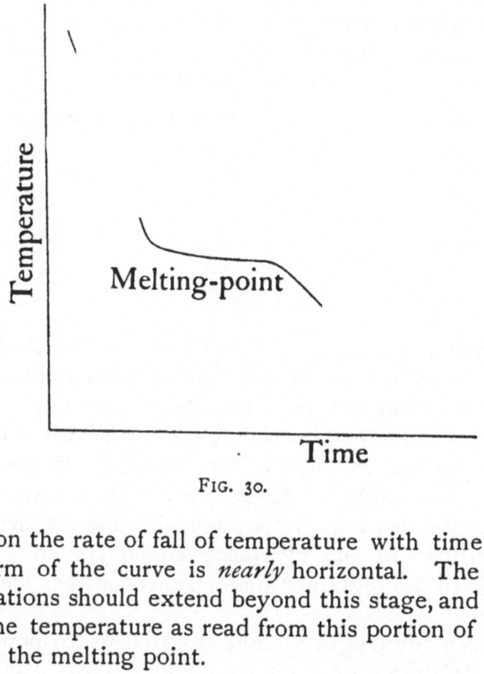

FIG. 30.

solidification the rate of fall of temperature with time is small, and the form of the curve is *nearly* horizontal. The range of the observations should extend beyond this stage, and the mean value of the temperature as read from this portion of the curve is taken as the melting point.

LINEAR EXPANSION OF A METAL.

Apparatus.—Tubes of brass, iron, glass, etc., about 1 metre long and say 5 or 6 mm. internal diameter, with flange attached near one end and fine pointer near the other end, stand with V-supports for above, boiler, cold water tap, microscope with micrometer eye-piece, thermometer.

Method.—Clamp one end of a tube to a fixed support, connect this end to the water tap, and send a stream of cold water through the tube till the temperature becomes constant. Focus the pointer on the scale of microscope eye-piece, and observe the position of the image of the pointer on this scale. Take the temperature of the water as it issues from the open end of the tube : let it be t_1° C.

Shut off the cold water supply and turn on the steam from the boiler, allowing this to pass through the tube until the temperature is raised to that of the steam, say t_2° C. During this rise of temperature the tube expands and the pointer moves across the scale of the microscope eye-piece. When the pointer comes to rest, observe the new position through the microscope.

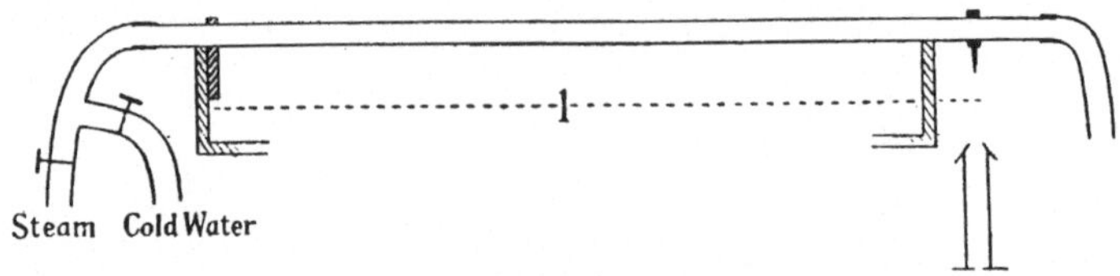

FIG. 31.

Again shut off the steam and turn on the cold water tap, and take readings as at the beginning of the experiment. Provided the temperature of the water is the same as originally, the position of the pointer should be as during the first reading. The expansion of the tube is thus obtained in eye-piece scale divisions. To calibrate these, focus the microscope on an accurately divided millimetre scale, and observe the number of scale divisions of the eye-piece equivalent to 1 mm. The expansion of the tube in scale divisions can then be reduced to millimetres.

Measure the length of tube from clamp to pointer. Let it be l_1 at t_1° C. and l_2 at t_2° C.

Then $l_1 = l_0 (1 + a t_1)$, in the usual notation,

and $l_2 = l_0 (1 + a t_2)$;

$$\therefore \quad \frac{l_2}{l_1} = \frac{1+at_2}{1+at_1} = 1+a(t_2-t_1) \text{ approximately.}$$

Knowing all the other quantities, a, the coefficient of linear expansion, may be calculated.

Make similar experiments with tubes of various materials.

MAXIMUM DENSITY OF WATER BY HOPE'S METHOD.

Apparatus.—Hope's apparatus consisting of a glass cylinder filled with water and surrounded about half-way up by a jacket which can be packed with a freezing mixture, and having two thermometers inserted through the side of the cylinder, one above and the other below the jacket.

Method.—Pack the jacket with a mixture of ice and salt, and take the readings of the two thermometers. It will be found that the lower thermometer, and thus the water in the lower portion of the cylinder, gradually falls in tempera-

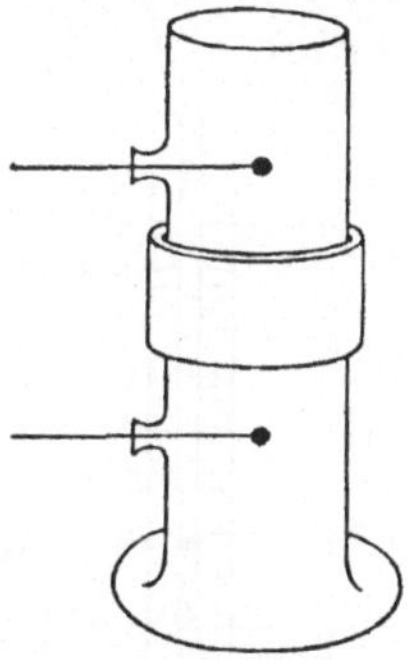

FIG. 32.

ture till about 4° C. is reached, while the temperature indicated by the upper thermometer changes *much* more slowly. When the lower thermometer indicates 4° C., the temperature of the upper one commences to fall, and does so till it reaches 0° C., while the temperature of the lower thermometer still remains at 4° C. This shows that the temperature of maximum density of water is 4° C.

MAXIMUM DENSITY OF WATER BY THE DILATOMETER METHOD.

Apparatus.—Dilatometer of the form in the diagram with a scale behind the stem, thermometer, beaker, powdered ice.

Method.—To correct for the expansion of glass of the dilatometer, a quantity of mercury of volume about one-seventh that of the bulb of the dilatometer should be put in the bulb, and the bulb then filled with water till the water reaches well up the stem.

Place the bulb in the beaker with some water, and take the

reading of the level of the water on the scale attached to the stem and the temperature of the water in the beaker. Put a small quantity of the powdered ice in the beaker, stir well till the beaker remains at a uniform temperature for some time. Then read the level of the water and the temperature. Repeat till the temperature reaches 1° C., taking small intervals of change of temperature in the region of 4° C. Now allow the water to rise in temperature, and take corresponding readings of the water level and temperature.

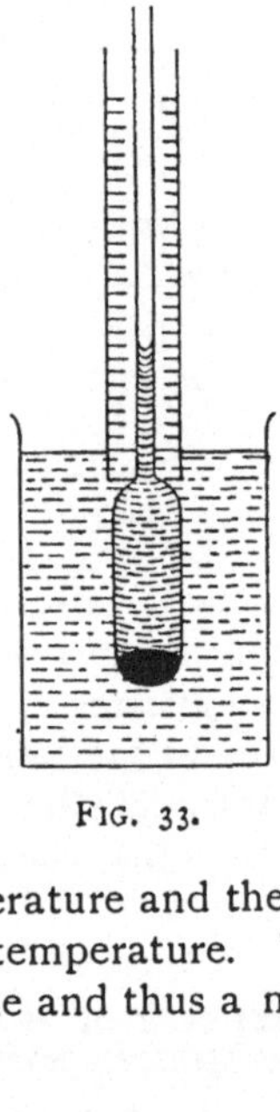

FIG. 33.

Plot a curve of temperature and the mean of the two readings of the level at that temperature. The graph will indicate a minimum in the volume and thus a maximum in the density of the water at 4° C.

COEFFICIENT OF THERMAL EXPANSION OF A LIQUID.

Coefficient of Expansion of the Liquid relative to that of the Vessel; or Apparent Coefficient of Expansion.

Apparatus.—Specific gravity bottle, beaker, thermometer, balance, weights.

Method.—Find the weight (W) of the specific gravity bottle empty and dry. Find also the weight (W_1) of the bottle full of water at a temperature of about 10°-15° C. This can be done by filling the bottle with water and immersing in a beaker of water for some time, till the temperature of the bottle and the water it contains has attained that of the water in the beaker. Let this temperature be t_1° C. Now immerse the bottle in the beaker again and raise the temperature of the water in the beaker till it boils. After the bottle has been in the boiling water for some minutes (long enough to enable the bottle and its contents to attain the temperature of the boiling water), take out the bottle by means of a thread attached to it, and dry and weigh. Let its weight be now W_2 and the temperature of the boiling water be t_2° C.

The weight of the liquid in the bottle is $W_1 - W$ at t_1° C., and $W_2 - W$ at t_2° C. If we neglect the change in volume of the bottle and let V be this volume, the volumes of unit weight of liquid at the two temperatures are $\frac{V}{W_1 - W}$ and $\frac{V}{W_2 - W}$ respectively, and therefore $\frac{V}{W_2 - W} = \frac{V}{W_1 - W}(1 + \beta\, \overline{t_2 - t_1})$ *approximately*, where β is the coefficient of apparent expansion of the liquid.

$$\text{Thus } \beta = \frac{W_1 - W_2}{(W_2 - W)(t_2 - t_1)}.$$

The absolute coefficient of expansion of the liquid is got by adding to the apparent coefficient of expansion, as determined above, the coefficient of cubical expansion of the glass.

RELATION OF THE VOLUME OF A GAS TO ITS TEMPERATURE AT CONSTANT PRESSURE.

Apparatus.—A uniform thin-walled glass tube sealed at one end and having a small thread of mercury enclosing a constant mass of air, thermometer, beaker, and heating apparatus; the tube is mounted on a scale reading to millimetres, so that the length of the air column can be measured.

Method.—The glass tube is put into a beaker of water, so that the whole air column is immersed. The water is heated up to about 80° C., and allowed to cool gradually. The zero reading of the sealed end of the tube is taken, and readings of the lower surface of the mercury thread are taken at intervals of 5° fall of temperature, the water being well stirred before each reading. The pressure is atmospheric + that due to the mercury thread (the tube being vertical), and is constant during the experiment. Readings are obtained of the volume V of the enclosed air—(these are proportional to the length, if the tube is uniform)—and of the corresponding temperature (t° C.), and V and t are tabulated.

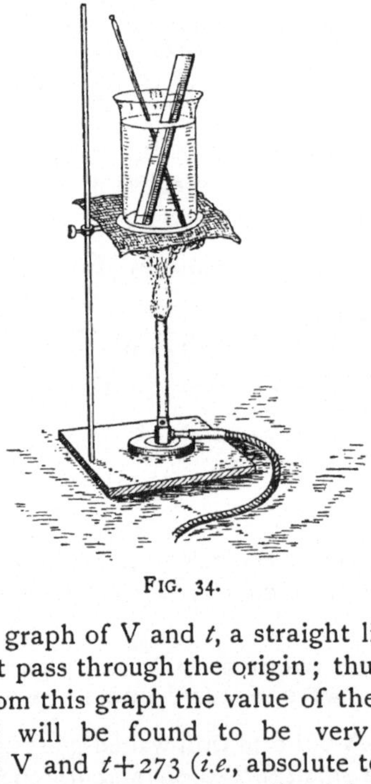

FIG. 34.

By plotting a graph of V and t, a straight line is got, which, however, does not pass through the origin; thus its equation is $t = m\,V + C$. From this graph the value of the constant C can be obtained; it will be found to be very nearly -273° C. Thus by plotting V and $t + 273$ (*i.e.*, absolute temperature T), a straight line is got passing through the origin, showing that the volume (V) varies directly as the absolute temperature (T). By calculation it can also be shown that the ratio V/T is constant.

RELATION OF THE PRESSURE OF A GAS TO ITS TEMPERATURE AT CONSTANT VOLUME.

Apparatus.—A glass bulb connected by a fine bore tube to the flexible U-tube arrangement CD which is filled with mercury and alongside of which is a metre scale to measure the heights of the mercury columns; vessel in which to immerse the bulb; thermometer.

Method.—Immerse the bulb B in the vessel containing water, and after stirring the water raise or lower the arm D until the level of the mercury in C is at some fixed mark P. Read the position of the mercury column in D and the temperature by the thermometer in B. Heat the water in stages of about 10°, and before taking readings take away the burner and stir the water in the vessel. Then again adjust the mercury level by raising D till the mercury in C stands at P. Read the level of mercury in D and the temperature of B. Take a series

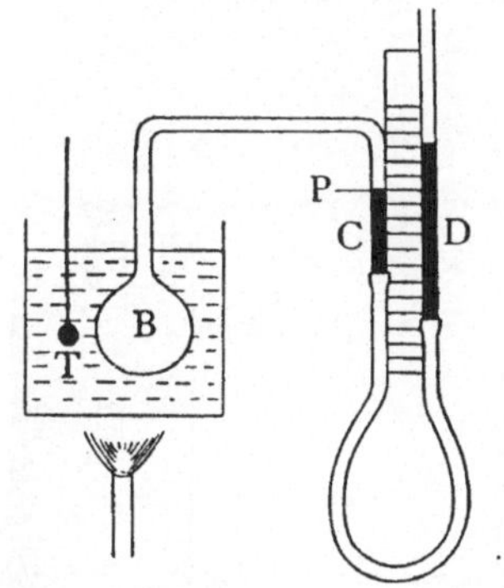

FIG. 35.

of such observations up to a temperature of about 80°. Allow the water in the vessel to cool down gradually, and take a similar series of readings of the temperature and the level of the mercury in D after adjustment of the level in C. Read the barometer during the course of the experiment. The total pressure of air in the bulb expressed in height of mercury is H + *h*, where H is the barometric height and *h* is the height of mercury in D above that in C.

Tabulate the results:—

Temperature $= t°$ C.	Height of Mercury in C.	Height of Mercury in D.	Difference of Level of C and D.	Total Pressure = P.

Plot a graph of temperature and pressure. This is a

straight line of the form $t = m\text{P} + \text{C}$. It will be found that the constant C is approximately $-273°$ C., so that if absolute temperatures T, *i.e.* $(t + 273)$ be used, then the graph of P and T is a straight line passing through the origin, and shows that the pressure is directly proportional to the absolute temperature.

By calculation it can also be shown that P/T is constant.

WATER EQUIVALENT OF A CALORIMETER AND SPECIFIC HEAT OF A METAL BY THE METHOD OF MIXTURES.

Apparatus.—Calorimeter with jacket, stirrer, thermometers (graduated in fifths of a degree), metal chips, heater for metal, balance, weights.

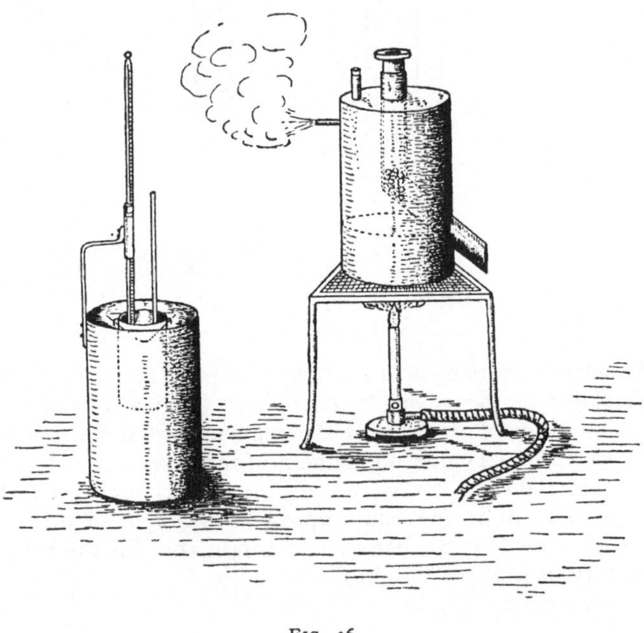

FIG. 36.

Method.—Weigh the empty calorimeter with stirrer; its mass say is m. Partly fill it with water at about ordinary room temperature (t_1); the mass of the water (M_1) may be found by weighing the calorimeter and the water and subtracting the weight of the calorimeter. Pour into the cold water in the calorimeter a mass of warm water (M_2) at temperature (t_2); stir the contents and observe the final temperature (t). The mass of the warm water added is now obtained by weighing as in the case of the cold water.

The water equivalent (w) of the calorimeter is calculated from the equation

$$(M_1+w)(t-t_1)=M_2(t_2-t),$$

which expresses the fact that the heat gained by the calorimeter, stirrer, and cold water is equal to that lost by the hot water.

To determine the specific heat of the metal proceed as above, except that instead of the hot water, the hot metal, which has been heated in the heater and its temperature taken, is added to the calorimeter. Let the mass of the metal, in this case, be M_2. The temperature (t_2) of the metal should be about 100° C.

During the time that the heat from the metal is being evenly distributed throughout the mixture, heat is being lost by radiation, etc., and accordingly a correction is made to allow for the consequent fall in temperature. This can be done by plotting a cooling curve from time-temperature readings. Make observations of the temperature of the mixture every quarter-minute for the first minute or two, and every minute after for several minutes, the water being kept well stirred. Choose two points on the latter portion of this graph, and find the fall in temperature per unit time; this may be taken as the rate of cooling corresponding to the mean of these two temperatures. Then by Newton's Law of Cooling, since the rate of cooling is zero at room temperature, we can find the rate of cooling at any temperature, and thus the rate of cooling at the temperatures with which we are concerned can be found, and a correction applied. (Strictly, Newton's Law applies only to loss by radiation; in this experiment, however, the other losses are small, consequently we may, without serious error, apply the law to correct for the total loss.)

An example will make the method clear. The observations will be somewhat as follows :—

Time (mins.).	Temp.	Mean.	Correction.	Total Correction.	Corrected Temperature.
0	16·4	...	...		
·25	24·5	20·45	·08/4=·02	·02	24·52
·5	26·8	25·65	·18/4=·045	·07	26·87
·75	27·5	27·15	·22/4=·055	·12	27·62
1	27·7	27·60	·22/4=·055	·18	27·88
2	27·6	27·65	·22	·40	28·00
3	27·4	27·50	·22	·62	28·02
4	27·18	27·29	·21	·83	28·01
5	26·9	27·04	·21	1·04	27·94
6	26·8	26·85	·20	1·24	28·04
7	26·55	26·68	·20	1·44	27·99
8	26·38	26·47	·20	1·64	28·02

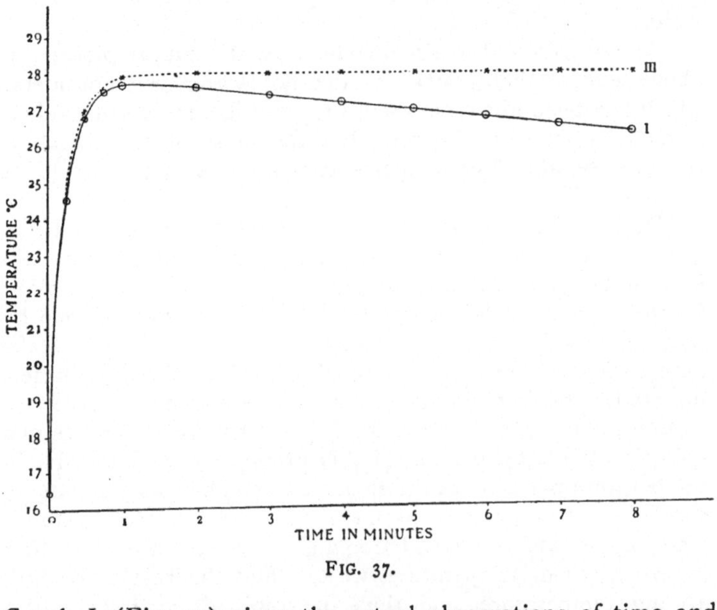

FIG. 37.

Graph I. (Fig. 37) gives the actual observations of time and temperature. From the fourth to the eighth minute the temperature fell ·80° C., *i.e.*, at the rate of ·2° C. per minute, the mean temperature being

$$\frac{27{\cdot}18+26{\cdot}38}{2} = 26{\cdot}8^\circ \text{ C.}$$

Thus, Graph II. (Fig. 38) can be drawn, the rate of cooling at 26·8° C. being ·2° C. per minute and being zero at room temperature, viz., 16·4° C.

The figures in column 4 are then got from Graph II., the rate of cooling per minute being the rate of cooling at the mean temperature. Column 5 is got by adding the corrections during the respective minutes, for it is clear that the fall in temperature is cumulative, *i.e.*, the fall in temperature in two successive minutes = fall in temperature during the first minute + fall in second minute.

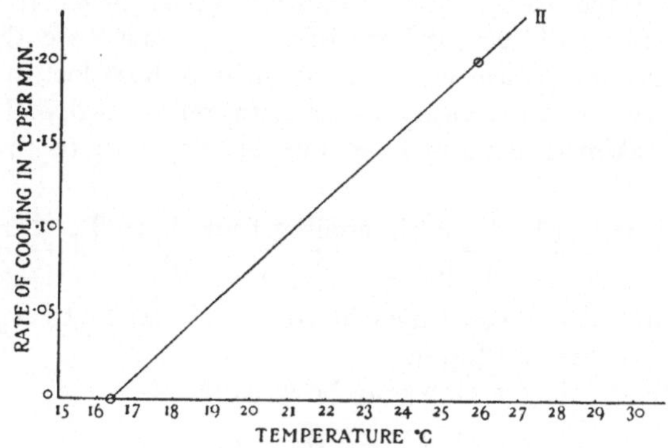

FIG. 38.

Column 6, which also appears as Graph III. (Fig. 37), is got by adding the total correction to the values of the temperature given in column 2.

The true final temperature (t) of the mixture is then 28·0° C., and equating heat lost to heat gained the specific heat (s) of the metal can be calculated from the formula

$$(M_1+w)(t-t_1)=M_2s(t_2-t).$$

SPECIFIC HEAT OF A LIQUID BY THE METHOD OF COOLING.

Apparatus.—Calorimeter and stirrer, thermometer, measuring glass, balance, weights.

Method.—Find the mass (m) of the empty calorimeter and fill it about two-thirds full of water at about 50° C., the volume of the water being measured by the measuring glass, from which it is poured into the calorimeter. Suspend the calorimeter inside its surrounding jacket, and, keeping the contents well stirred, note the times when the temperature of the calorimeter is 50°, 49°, 48°, etc. Find the mass of the water in the calorimeter by weighing calorimeter and contents.

Do the same for the liquid whose specific heat is required, care being taken that the volume of liquid used is approximately the same as that of the water.

Since the amount of heat radiated from the surface of the calorimeter depends only on the difference of temperatures of the calorimeter and enclosure and on the nature and area of the surface from which radiation takes place, it is seen that the rates of loss of heat in the two cases are the same at the same temperatures. The quantity of heat lost in cooling between two given temperatures is therefore proportional to the time taken in cooling from one temperature to the other, that is,

$$\frac{\text{Heat lost by liquid in cooling from } T_1 \text{ to } T_2}{\text{,, \quad ,, \quad water ,, \quad ,, \quad } T_1 \text{ to } T_2} = \frac{t_2}{t_1},$$

where t_1 and t_2 are the times of cooling from T_1 to T_2 in the two cases, water and liquid.

But the above fraction may be written

$$\frac{M_2 s(T_1 - T_2)}{M_1(T_1 - T_2)} = \frac{t_2}{t_1},$$

where M_1 and M_2 are the masses of water and liquid respectively.

Taking the water equivalent of the calorimeter as w, we have more correctly

$$\frac{M_2 s + w}{M_1 + w} = \frac{t_2}{t_1}.$$

We then determine s, the specific heat of the liquid, knowing all the other quantities.

LATENT HEAT OF STEAM OR LATENT HEAT OF VAPORISATION OF WATER.

Apparatus.—Calorimeter, boiler to generate steam, balance, weights, thermometer.

Method.—Find the mass (m) of the empty calorimeter and stirrer and part fill with water at ordinary room temperature.

Find by weighing the calorimeter, stirrer, and water the mass (M_1) of the water in it; take the temperature of the water ($t°$). Pass steam into the calorimeter from the boiler till the temperature of the calorimeter and contents is about 30° C. Find the final temperature of the calorimeter and contents (T°) by the thermometer, and also the mass M_2 of steam condensed by weighing the calorimeter with the water and condensed steam.

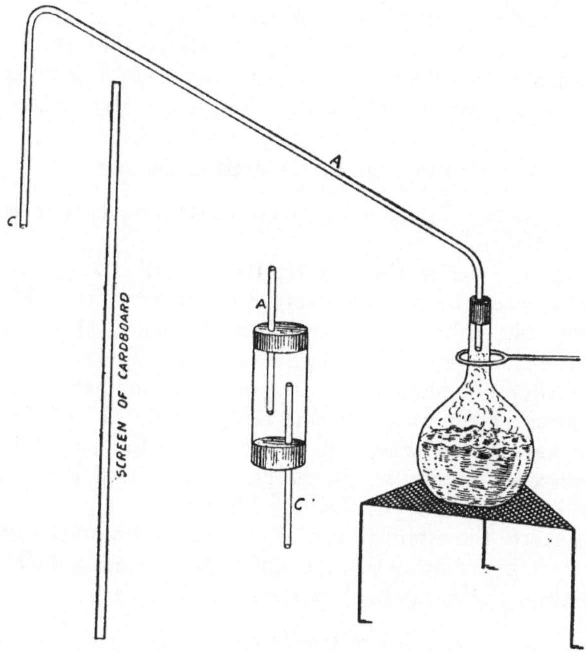

FIG. 39.

The latent heat of steam can be calculated from the formula—Heat gained by water and calorimeter = Heat lost by steam in condensation + Heat lost in further cooling.

$$(M_1+w)(T-t)=M_2L+M_2(100-T),$$

where w is the water equivalent of the calorimeter and stirrer—that is, mass of calorimeter and stirrer × specific heat of the material—and L is the latent heat of steam, here assumed to be at 100° C.

To minimise condensation in the conducting pipe of the boiler, the pipe should be short and surrounded by some non-conducting material, *e.g.*, cotton wool, or there should be a water trap in this portion. To ensure greater accuracy in the experiment without the complication of correcting for the loss of heat by cooling, the water in the calorimeter should originally be a few degrees, say 5°, below the temperature of the air. Steam should then be passed until the temperature is about the same number of degrees above that of the room. The warming in the first half of the experiment is thus approximately balanced by the cooling in the second half.

LATENT HEAT OF FUSION OF ICE.

Apparatus.—Calorimeter, pieces of ice, filter paper, balance, weights, thermometer.

Method.—Weigh the calorimeter and fill about two-thirds full of water at a temperature of about 30°. Find by weighing the calorimeter the mass of water (M_1) in it, and then take the temperature of the water (t). Quickly drop into the calorimeter some clear pieces of ice which have been dried with filter paper, and note the temperature (T) of the water in the calorimeter when all the ice is melted and the water well stirred. Finally weigh the calorimeter and contents, and thus find the mass M_2 of the ice added.

Calculate the latent heat of fusion of ice from the formula—
Heat lost by calorimeter and contents = Heat gained by ice in melting + Heat gained by ice after melting.

$$(M_1+w)(t-T)=M_2L+M_2T,$$

where w is the water equivalent of the calorimeter and stirrer (*i.e.*, mass of these × specific heat of their material) and L is the latent heat of fusion of ice.

LATENT HEAT OF FUSION OF A SOLID.

Apparatus.—Calorimeter, paraffin wax, vessel with hot water, thermometer, balance, and weights.

Method.—Find the mass (m) of the empty calorimeter

and add a few grams of paraffin wax and weigh again, and thus find the mass of the wax (M_1). Take the temperature (t) of the calorimeter and wax, and add to the calorimeter with the wax sufficient hot water (at temperature t_1) to melt the wax completely; stir and find the final temperature (T) of the calorimeter, wax, and water. Find the mass M_2 of the hot water added to the calorimeter, and calculate the latent heat of fusion (L) of the wax from the expression

$$M_1L+(w+M_1s)(T-t)=M_2(t_1-T),$$

where w = water equivalent of calorimeter, stirrer, and thermometer, and s = specific heat of the paraffin wax, here assumed to be the same in the solid and liquid states.

HYGROMETRY.

Dew Point.—The dew point may be obtained by hygrometers of various forms. From the dew point and the temperature of the air, the hygrometric state of the atmosphere may be got from tables of saturation vapour pressure.

Dines's Hygrometer.

Turn tap and allow the cold water from the vessel to flow slowly past the thermometer and glass plate until dew begins

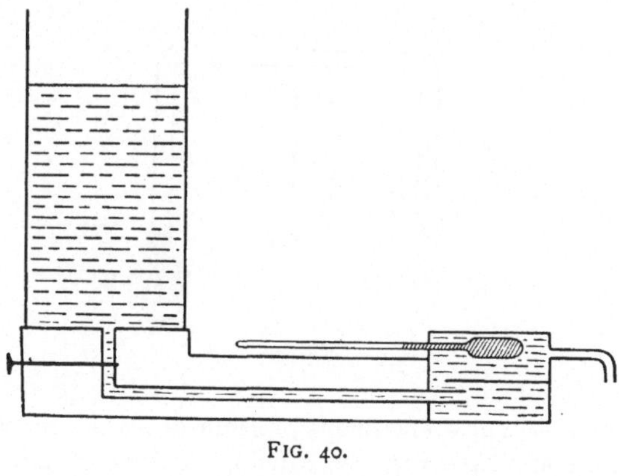

FIG. 40.

to appear on the surface of the plate. Read the temperature at which this occurs. Shut tap and observe the deposit of dew on the plate until it disappears. Read the temperature. Take the mean of the two temperatures at which dew appears and disappears. Repeat the experiment several times and take the mean of the results obtained.

This gives us an approximate value for the dew point, that is, the temperature at which the actual pressure of aqueous vapour in the atmosphere is the saturation vapour pressure. From saturation vapour pressure tables find the saturation vapour pressure for this particular temperature. Find also the saturation vapour pressure for the actual temperature of the air as given by a thermometer outside.

Then relative humidity of atmosphere

$$= \frac{\text{saturated vapour pressure at dew point}}{\text{saturated vapour pressure at temperature of air}}.$$

Daniell's Hygrometer.

Apparatus.—Hygrometer, ether. (This hygrometer consists of two glass bulbs connected together by a wide tube bent twice at right angles as shown in the figure. Bulb A contains ether and is coated on its outer surface by a thin layer of polished silver or gold. Bulb B is covered with muslin. Thermometers are situated inside A and outside between A and B.)

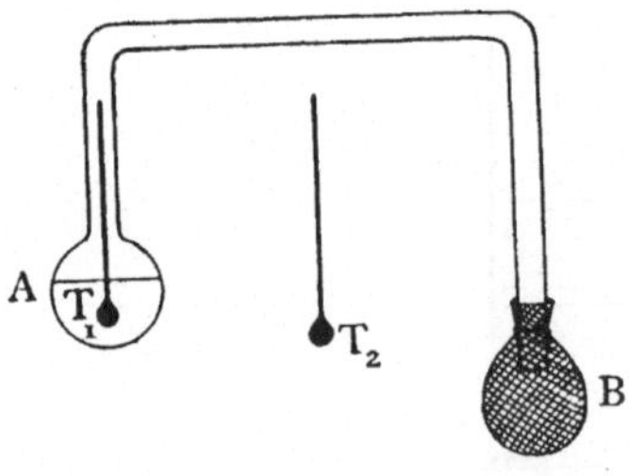

FIG. 41.

Method.—Pour a few drops of ether on to the muslin round B. This evaporates rapidly, causing cooling of the bulb and ether vapour inside, consequently ether vapour flows from bulb A to bulb B and leaves the space in A unsaturated. Ether inside evaporates and causes cooling of A and of the polished metal on the outside. The muslin on B is kept moist until, when a certain temperature is reached, a deposit of dew is formed on the metal surface. Note this temperature given by the thermometer T_1. After allowing the ether on the muslin to disappear by evaporation, the temperature of A rises, and the dew disappears. Note the temperature indicated by T_1 at the instant of disappearance. Take the mean of these two temperatures. Repeat the experiment a number of times, and take the mean of all these temperatures as the dew point. Thermometer T_2 gives the air temperature. This should also be read.

From the dew point and the air temperature, the humidity of the atmosphere may be obtained as explained from Saturated Vapour Pressure Tables.

Regnault's Hygrometer.

In this form of hygrometer a test tube has a covering of silver foil on its outer surface. This is cooled down by passing a stream of air through ether contained in the tube. The rapid evaporation causes cooling of the liquid, and consequently of the tube and silver. As in Dines's hygrometer, the temperature of the silver at which dew begins to deposit is given by the thermometer enclosed.

An observation of the temperature is also taken when the dew disappears after the stream of air has been stopped. The mean of these two temperatures is taken. The experiment is repeated a number of times and the mean of the observations obtained.

The temperature of the air is given by a thermometer contained in a similar tube, which is, however, otherwise empty and is not artificially cooled. Proceed as described in the last experiment to determine the relative humidity of the atmosphere from Saturated Vapour Pressure Tables.

Dry and Wet Bulb Hygrometer.

Two direct readings of the temperatures as given by the ordinary (dry bulb) thermometer and by the thermometer (wet bulb), from which evaporation is continually taking place, are sufficient to determine with the help of tables the humidity of the atmosphere. For if we multiply the difference between these two readings by Glaisher's Factor (f, in Table, p. 112) we can determine the dew point from the formula

$$\frac{\text{Temperature of air} - \text{dew point temperature}}{\text{Temperature of dry bulb} - \text{temperature of wet bulb}} = f.$$

From the temperature of the air and the dew point, determine the humidity of the atmosphere as before by use of Saturated Vapour Pressure Tables.

ELECTRICITY AND MAGNETISM.

STATICAL METHOD OF COMPARISON OF MAGNETIC MOMENTS.

Apparatus.—Deflection magnetometer, bar magnets, metre scale.

Method.—(*a*) Position of Gauss (end on).

Place one of the magnets to be compared in an east-west position, with its axis passing through the centre of a small magnetometer needle (see Fig. 42); note deflection θ_1 of the

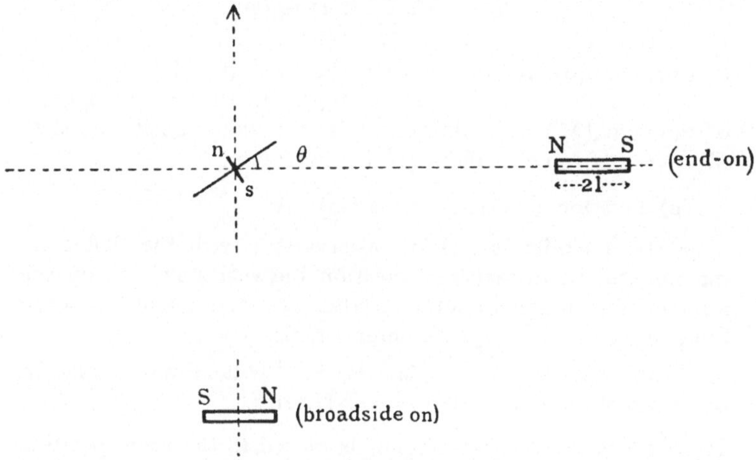

FIG. 42.

magnetometer needle, reading both ends and taking the mean value. Measure also the distance d between the centres of the deflecting and deflected magnets. Reverse the deflecting magnet; read deflection θ_2. Place the deflecting magnet at an equal distance on the other side of the magnetometer needle

E

and take similar readings. Finally, obtain the mean deflection for the distance d between the centres of the deflecting and deflected magnets.

If H be the horizontal component of the earth's magnetic field, M the magnetic moment of the deflecting magnet, $2l$ the length of this magnet, then approximately

$$M = \frac{(d^2-l^2)^2}{2d} H \tan\theta;$$

as a first approximation we may take $M=\frac{1}{2}d^3 \times H \tan\theta$.

Take a similar series of readings with the second magnet of moment (M′) with its centre in the same positions, then $M'=\frac{1}{2}d^3 \times H \tan\theta'$. Then $M/M'=\tan\theta/\tan\theta'$.

(*b*) Position of Gauss (broadside on).

Make a similar series of observations with the deflecting magnet still in an east-west position but with its centre on the axis of the magnetometer needle, *i.e.*, due north or south (magnetic) of the magnetometer needle.

In this case $M=(d^2+l^2)^{\frac{3}{2}} H \tan\theta$ approximately,
or, as a first approximation $M=d^3 H \tan\theta$.

If the second deflecting magnet is placed in the same position producing a deflection θ', $M/M'=\tan\theta/\tan\theta'$.

Method of Equal Deflections.—If the two magnets to be compared be placed in either the (*a*) or (*b*) position, but at distances arranged so as to give equal deflections of the magnetometer needle, then from the above formula it can be seen that $M/M'=d^3/d'^3$ to a first approximation.

Null Method.—If the two deflecting magnets be placed in position (either (*a*) or (*b*)) at the same time, but in directions so arranged that the separate deflections are in opposite directions, one may be made to neutralise the effect of the other so that there is no deflection of the magnetometer needle.

Then position (*a*) $2M/d^3=2M'/d'^3$,
or position (*b*) $M/d^3=M'/d'^3$;
thus in either case $M/M'=d^3/d'^3$.

STATICAL METHOD OF COMPARISON OF INTENSITIES OF MAGNETIC FIELDS.

Apparatus.—Uniform fields of strengths H_1 and H_2,* bar magnet, deflection magnetometer.

Method.—If the bar magnet is placed in the field H_1 and then in H_2, in either of the two positions of Gauss relative to the magnet of the magnetometer, then

$$d_1^3 H_1 \tan\theta_1 = d_2^3 H_2 \tan\theta_2 ;$$

$$\text{if} \quad d_1 = d_2 \text{ then } \frac{H_1}{H_2} = \frac{\tan\theta_2}{\tan\theta_1}.$$

At the place where the field H_1 exists, place the bar magnet and magnetometer in one of the Gauss positions (say the (*a*) position) and find θ_1; keeping d_1 the same, reverse the magnet and again read the magnetometer. Repeat these two observations when the bar magnet is on the opposite side of the magnetometer and again having d_1 the same as before. Do the same where the field is H_2, whence the fields can be compared.

VIBRATION METHOD OF COMPARISON OF MAGNETIC MOMENTS.

Apparatus.—Magnets to be compared, stirrup for suspension, supported by silk fibre, stop-watch.

Method.—(*a*) If a bar magnet of magnetic moment M and moment of inertia I about the axis of suspension be caused to execute small angular oscillations about its axis of suspension in a uniform magnetic field of strength H, then the time of a complete oscillation is given by

$$T = 2\pi\sqrt{\frac{I}{MH}}.$$

After removing the torsion of the fibre by suspending from it a weight equal to that of the magnet, allow the first magnet to oscillate in the earth's magnetic field, and find the time of oscillation. The oscillations of the magnet may be started by bringing near it another magnet, care being taken that the subsequent motion is one of torsional oscillation only; this latter magnet is then removed to a distance.

* It will usually be found that iron girders, pipes, etc., so modify the field, that in an ordinary building or even in one room fields of very different strength may be found, each sufficiently uniform for the purpose of this experiment.

Repeat for the second magnet.

$$\text{Then} \quad \frac{M_1}{M_2} = \frac{I_1}{I_2} \cdot \frac{T_2^2}{T_1^2}.$$

The moments of inertia I_1 and I_2 can be calculated if the magnets have a simple geometrical form ; they are equal if the masses and dimensions are the same.

(*b*) In the following method a knowledge of the moments of inertia is not required :—

Suspend the two magnets so as to form one oscillating system with their like poles (1) in the same direction, and (2) in the opposite direction, and find T_1 and T_2 the periods of vibrations in the two cases. Care must be taken that the positions are exactly the same in the two cases, except for the reversed polarity of one.

In the first case the magnetic moments are added, and in the second subtracted, while the moment of inertia of the combination is the same in the two cases, thus,

$$\frac{M_1+M_2}{M_1-M_2} = \frac{T_2^2}{T_1^2}, \text{ whence}$$

$$\frac{M_1}{M_2} = \frac{T_2^2+T_1^2}{T_2^2-T_1^2}.$$

VIBRATION METHOD OF COMPARISON OF INTENSITIES OF MAGNETIC FIELDS.

Apparatus.—Vibration magnetometer, fields to be compared, stop-watch.

Method.—Deflect the magnet so as to produce oscillations through a small angle in the field of strength H_1, and find the time of oscillation T_1 from the time of a large number of complete oscillations.

Repeat in the second field of strength H_2. Let the time of one complete oscillation be T_2.

$$\text{Then as} \quad T_1 = 2\pi\sqrt{\frac{I}{MH_1}},$$

$$\text{and} \quad T_2 = 2\pi\sqrt{\frac{I}{MH_2}};$$

$$\therefore \frac{H_1}{H_2} = \frac{T_2^2}{T_1^2}.$$

MEASUREMENT OF THE ABSOLUTE VALUE OF THE HORIZONTAL COMPONENT OF THE EARTH'S MAGNETIC FIELD.

Apparatus.—Deflection magnetometer, bar magnet, stirrup for suspension, metre scale, stop-watch.

Method.—As in experiment, p. 65, find by a statical method the value of M/H. As in experiment, p. 67, find by a vibration method the value of MH. In this case the moment of inertia of the bar magnet requires to be known; this may be got by calculation if the magnet is of simple geometrical shape, or by experiment (see p. 42).

From M/H and MH, H may be determined and so also may M.

MAPPING OF MAGNETIC FIELDS.

Apparatus.—Bar magnet, very short compass needle, drawing board, and paper.

Method.—A line of magnetic force is a line the tangent to which at each point gives the direction of the magnetic force at that point.

If a very short magnetic needle be free to move about its centre of gravity, it comes to rest with its axis tangential to a line of force. A short pivoted compass needle is, however, sufficient to show the direction of the horizontal component of the field at any point. Dealing only with the horizontal field, it is possible to map out the lines of magnetic force due to the earth's field, combined with that of a bar magnet say, thus—

Fix the paper on the drawing board, and place the bar magnet on the paper so that its magnetic axis is parallel to magnetic meridian, and (*a*) its north pole (*b*) its south pole, pointing north respectively. Place the compass needle PQ in the field, and mark the positions of its ends P and Q. Then move it till the end P coincides with the mark Q, and mark the position Q_1 of its other end, and so on till the line of force is completed; the curve PQQ_1Q_2 . . . is then drawn. By starting from different positions and repeating the process, the whole field in one plane may be mapped out.

Do the same in the case of the magnet lying at right angles to the magnetic meridian.

In the diagram giving the lines of force there will be certain regions which show *neutral* points, *i.e.*, regions in which the horizontal fields of the magnet and the earth neutralise. Indicate these positions on your diagrams. Such a region is shown below at R. Knowing H, the field of the magnet at the neutral point is known; from this and the distance from the magnet, the magnetic moment M can be determined.

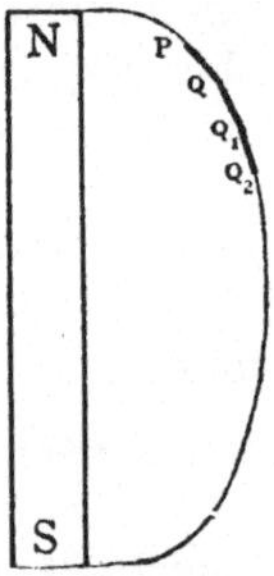

FIG. 43.

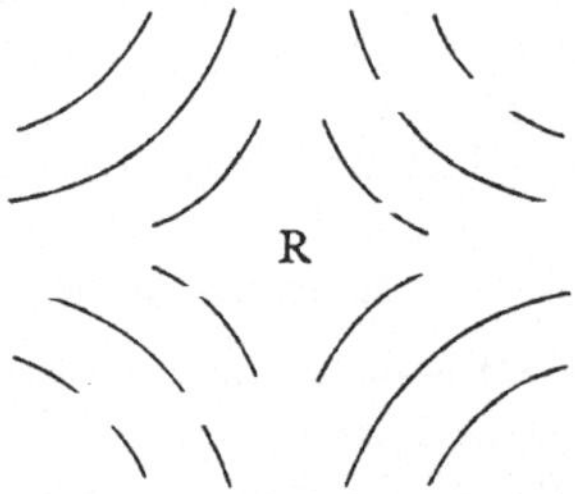

FIG. 44.

By the use of iron filings the fields may be mapped more quickly, but this method is not suitable for weak fields.

THE TANGENT GALVANOMETER

In all cases in which a tangent galvanometer is used, attention must be paid to certain necessary adjustments. The following is a brief description of a simple instrument and the method of using it. The instrument (Fig. 45) consists of a vertical coil of wire, at the centre of which is placed a very *small* magnet supported on a pivot, and capable of rotating in a horizontal plane with the minimum of friction. At right angles to the magnet and lying horizontally when the magnet is on its pivot is fixed a light pointer, and underneath the magnet and pointer is a circular scale graduated in degrees, which enables the angle of deflection of the magnet to be determined by reading the pointer. This scale sometimes consists of a mirror with the scale on its surface, so that parallax errors may be avoided.

The coil may consist of several independent coils, each having a different number of turns and a different resistance; by an arrangement of terminals at the base of the instrument any of these coils may be used.

When the plane of the coil is in the magnetic meridian and a current is passed through the coil, the measure of this current is given by $C = \frac{Hr}{2\pi n} \tan\theta$ in electromagnetic C.G.S. units, where

r = mean radius of coil in centimetres.
n = number of turns.
H = horizontal intensity of earth's magnetic field in C.G.S. units,
and θ = angle of deflection of the magnet.

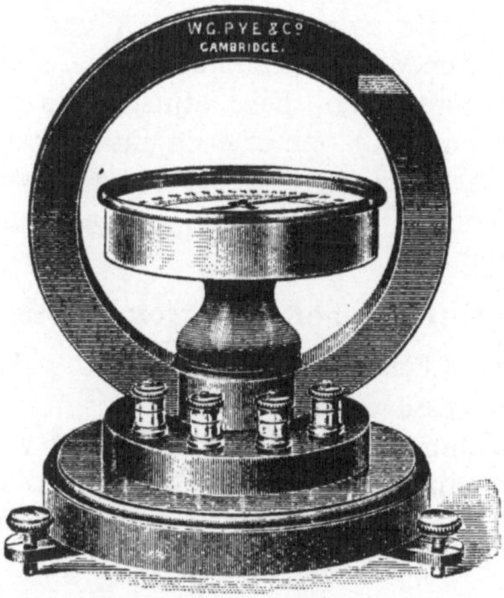

FIG. 45.

It will be noticed that since r and n are constants, and H is constant at one place, then currents are simply proportional to the tangents of the deflections; or $C = k \tan\theta$, where k is the "reduction factor" or "working constant" of the galvanometer.

Adjustments.—The detailed method of adjusting the instrument depends upon the particular pattern used. The following, however, describes the essential features. Rotate the coil about its vertical axis until its plane is approximately in the magnetic meridian. This is the case when the plane of the coil contains the axis of the needle. The central plane of the coil should then pass through the 90° divisions on the horizontal scale below the needle, and the pointer attached to and at right angles to the needle should indicate zero deflection. Level the base by means of a spirit-level and levelling screws, and make any slight readjustments of the plane of coil or position of horizontal scale that may be necessary. To test the accuracy of adjustment and correct inaccuracy if such is found, send a moderate current through the instrument by means of a cell and resistance, and read the deflections of the two ends of the pointer attached to the magnet; reverse the current, and again read the deflections obtained. If the means of the two pairs of readings differ appreciably, say by more than a degree or two, turn the coil and horizontal scale together round a vertical axis till this divergence is reduced to a minimum. In reading a deflection on the galvanometer when these adjustments have been made, two readings (θ_1 and θ_2), one of each end of the pointer, should be taken, and then the current reversed and two other readings (θ_3 and θ_4) made. The deflection produced is the mean of θ_1, θ_2, θ_3, and θ_4.

STUDY OF POLARISATION OF CELLS.

Apparatus.—Leclanché cell, tangent galvanometer, key, resistance (if necessary), stop-watch.

Method.—Connect the cell (B) in series with the galvanometer (G), resistance (R), and key (K) as in the diagram. The adjustment of the galvanometer having been made, the key K

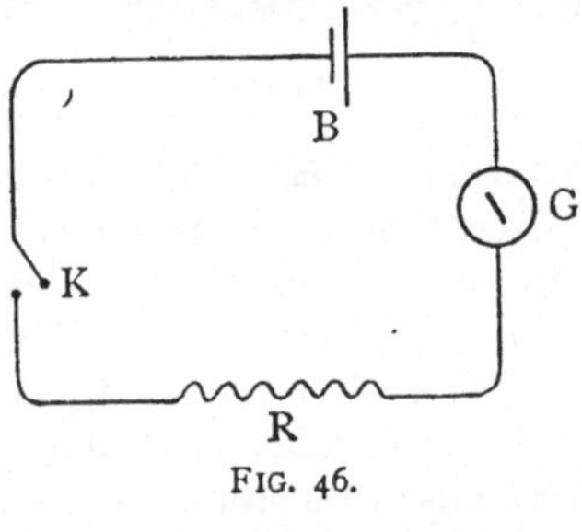

FIG. 46.

is closed at a given instant which is noted, and as soon as the needle comes to rest the current as indicated by the deflection —(tangent of angle in case of tangent galvanometer)—is observed.

A series of readings of current and time is taken at intervals of say a minute for about ten minutes—the most suitable time depends upon the apparatus used—and a graph of current and time is plotted. To observe the recovery of the cell—the depolarisation—open the key K, and after definite intervals close K for as short a time as possible in order to read the current value and open again. (The periods for which K is closed should be short compared with those for which it is open.)

Repeat with various types of cells.

RESISTANCE BY SUBSTITUTION METHOD.

Apparatus.—Battery of constant E.M.F., galvanometer, resistance to be measured, two-way key, resistance box.

Method.—Arrange the apparatus as in the diagram, where B is the battery, G the galvanometer, x and R the unknown and known resistances respectively, and K the key.

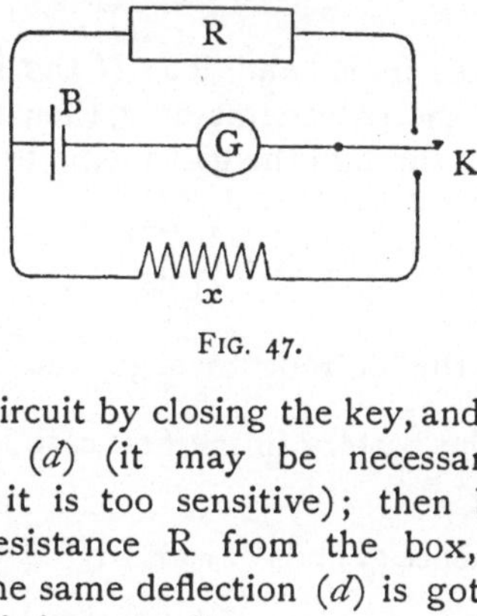

FIG. 47.

Put x in the circuit by closing the key, and find the galvanometer deflection (d) (it may be necessary to shunt the galvanometer if it is too sensitive); then by means of the key put in a resistance R from the box, and adjust this resistance until the same deflection (d) is got on the galvanometer. When this is so, $x = R$.

RESISTANCE OF A BATTERY BY THOMSON'S EQUAL DEFLECTION METHOD.

Apparatus.—Galvanometer, two resistance boxes, battery of constant E.M.F., keys.

Method.—Connect up as in the figure. Close the keys K_1 and K_2 and note the deflection (d) of the galvanometer, the resistances in the boxes R_1 and R_2 being adjusted so as to get a convenient deflection. Let the values of these resistances be r_1 and r_2. Then open K_2 and adjust the resistance in R_1 till the same deflection (d) is obtained, the value of the resistance being r_3. Repeat the above observations for

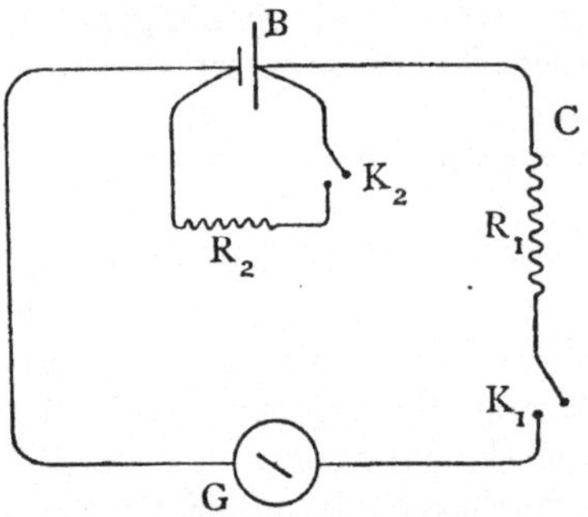

FIG. 48.

different resistances in R_1 and R_2. If the battery resistance be b, and that of the galvanometer g, then from the equality of the two currents through the galvanometer we have

$$b = r_2 \cdot \frac{r_3 - r_1}{r_1 + g}.$$

For if c be the current through the galvanometer the current through the battery in the first case is $c \times \frac{r_2 + r_1 + g}{r_2}$,

$\therefore$ potential difference at ends of wire BK_1GB is $e - \frac{r_2 + r_1 + g}{r_2} cb$

and current through BK_1GB is $\dfrac{e - \frac{r_2 + r_1 + g}{r_2} cb}{r_1 + g} = c.$

In the second case the current is $\dfrac{e - cb}{r_3 + g} = c,$

$\therefore$ subtracting numerators and denominators $cb \dfrac{\frac{r_1 + g}{r_2}}{r_3 - r_1} = c,$

$$\therefore \quad \frac{b}{r_2} = \frac{r_3 - r_1}{r_1 + g}.$$

RESISTANCE OF A GALVANOMETER BY THE EQUAL DEFLECTION METHOD.

Apparatus.—Galvanometer, two resistance boxes, battery of constant E.M.F. and of low resistance, keys.

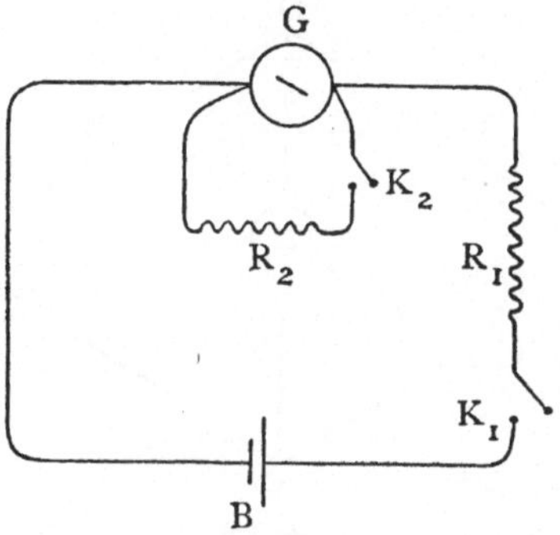

FIG. 49.

Method.—Connect up as in the figure. Close the keys K_1 and K_2 and note the deflection (d) of the galvanometer, the resistances in the boxes R_1 and R_2 being adjusted so as to get a convenient deflection. Let the values of these resistances be r_1 and r_2. Then open K_2 and adjust the resistances in R_1 till the same deflection (d) is obtained, the value of the resistance being r_3. Repeat the above observations for different resistances in R_1 and R_2. If the galvanometer resistance be g, then from the equality of the two currents through the galvanometer we have

$$\frac{e}{r_1 + \dfrac{r_2 g}{r_2+g}} \cdot \frac{r_2}{r_2+g} = \frac{e}{r_3+g},$$

where e is the E.M.F. of the battery the resistance of which is assumed to be negligible.

From the above we have $\quad g = \dfrac{r_2(r_3 - r_1)}{r_1}.$

THE WHEATSTONE BRIDGE.

By means of this method devised by Wheatstone, resistances can be accurately and easily measured. The arrangement is represented diagrammatically in the figure where E is a battery, G a galvanometer, and AB, BC, AD, DC, represent resistances. It can be shown that the condition that B and D are at the same potential—that is, that no current flows through the

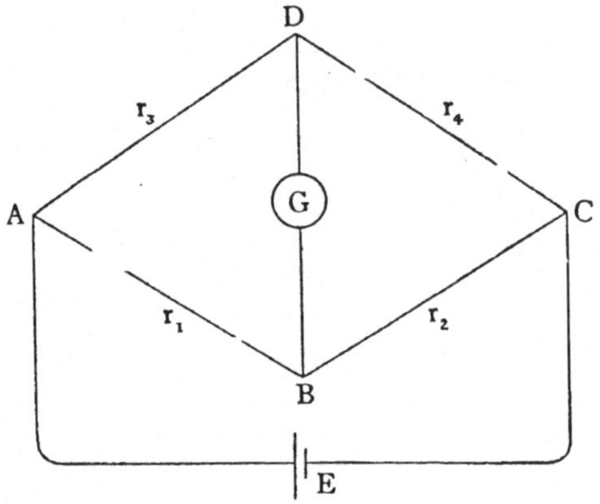

FIG. 50.

galvanometer—is that $r_1/r_2=r_3/r_4$; hence if only one of these magnitudes r_1, r_2, r_3, r_4 is unknown, its value can be calculated from the above equation. This scheme is adopted practically in the slide wire bridge and the post-office box.

THE SLIDE BRIDGE.

Electrical Resistance of a Wire and of Wires in (*a*) Series (*b*) Parallel.

Apparatus.—Metre bridge, sensitive galvanometer, battery, known resistance, wire whose resistance is to be measured.

Method.—Set up the apparatus as in the diagram, where B is the battery, K a key, G the galvanometer, and R and x are the known and unknown resistances respectively. Put the battery in circuit by means of the key K, and move the tapper of the bridge till a point N is found on the bridge wire such that when the galvanometer is in circuit no deflection occurs. When this is the case then

$$\frac{x}{R} = \frac{MN}{NL},$$

from which x, the unknown resistance, can be calculated.

For wires in series take a length of wire 2, 3 . . . times that of the original wire, and show that the resistance is increased into the ratio 2, 3 . . . times that of the first wire. This shows that the resistances of 2, 3 . . . uniform wires of equal length are proportional to the total length.

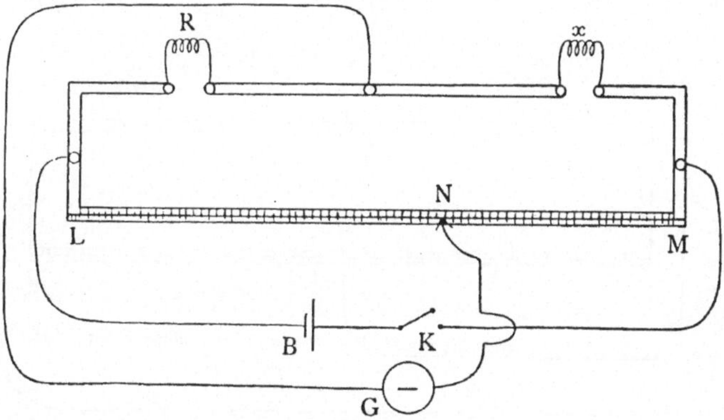

FIG. 51.

For wires in parallel use the frame shown in Fig. 52, and connect PQ to the gap of the bridge to which the wire of unknown resistance is joined, and find the resistance of 6, 5

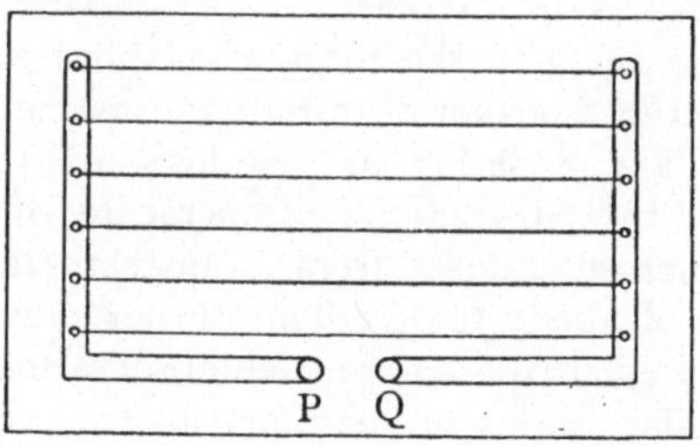

FIG. 52.

. . . 2 wires in parallel, and finally that of one wire by removing the wires successively. All the wires used must be of the same material, length, and cross section.

Show by a graph or by calculation that the resistance of the wires multiplied by their number is constant.

Notes.—Unless the leads connecting the wires, the known resistance, etc., to the bridge have a negligible resistance, it may be necessary to take account of their resistance in proving the above laws.

It is possible by adopting the arrangement given below, to dispense with the key K, and to use instead the bridge tapper.

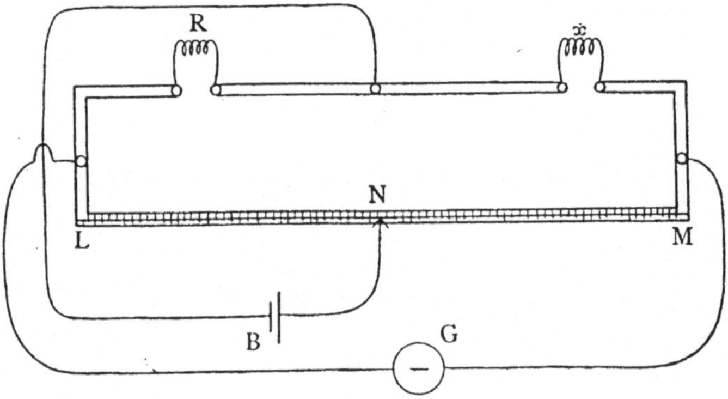

FIG. 53.

VARIATION OF RESISTANCE OF A WIRE WITH TEMPERATURE.

Apparatus.—As in the case of the resistance of a wire (p. 76), the wire being in the form of a spiral and capable of being immersed in a beaker of water; thermometer.

Method.—Take a series of readings of resistance and temperature of the wire as the temperature of the water in which it is immersed is raised from room temperature to about 90° C. by steps of about 10° C. The burner must be removed and the water well stirred before each observation.

Take a similar series of readings as the water and wire gradually cool.

Plot a graph of resistance and temperature, which with the range of temperature used will be a straight line whose equation is $R_t = mt + R_o$,

where R_t = resistance at temperature $t°$ C. and

R_o = resistance at temperature 0° C.

R_o can be got from the graph by a slight extrapolation.

Show also that $$\frac{R_t - R_o}{t} = m \text{ (a constant)}$$

or transforming to give R_t in terms of R_o

$$R_t = R_o + mt$$
$$= R_o (1 + \frac{m}{R_o} t)$$
$$= R_o (1 + at) \text{ where } a \text{ is a constant.}$$

COMPARISON OF ELECTROMOTIVE FORCES BY WIEDEMANN'S METHOD (Sum and Difference Method).

Apparatus.—Galvanometer, adjustable resistance, reversing key, two cells whose E.M.F.'s are to be compared.

Method.—(The method is unsuitable for cells in which appreciable polarisation is produced under the conditions of the experiment.)

Connect the cells E_1 and E_2 in series by means of the key K with the galvanometer G and the adjustable resistance R_1 as in the figure. Resistance R_1 must be of magnitude appropriate to the particular galvanometer used.

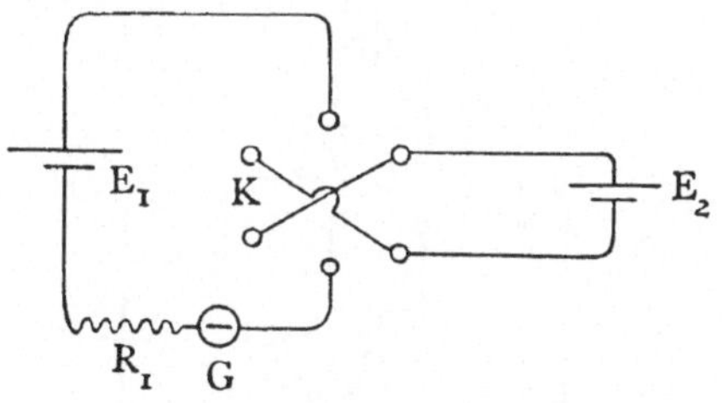

FIG. 54.

The first operation consists in having E_1 and E_2 ($E_1 > E_2$) so that their effects are additive, *i.e.*, the E.M.F.'s assist one another, which can be done by means of the key K. The deflection (d_1) of the galvanometer is noted. Now throw the key over so that the effects of E_1 and E_2 are subtractive, *i.e.*, the E.M.F.'s oppose one another, and again note the deflection (d_2) of the galvanometer. If the total resistance in the circuit be denoted by R, then the currents in the two cases C_1 and C_2, which are proportional to d_1 and d_2 (tan d_1 and tan d_2 must be employed instead of d_1 and d_2 if a tangent galvanometer is used) are

$$C_1 = \frac{E_1+E_2}{R} = kd_1; \qquad C_2 = \frac{E_1-E_2}{R} = kd_2$$

$$\therefore \frac{E_1+E_2}{E_1-E_2} = \frac{d_1}{d_2}; \qquad \text{whence } \frac{E_1}{E_2} = \frac{d_1+d_2}{d_1-d_2}.$$

Repeat the two above operations for different values of R_1, and finally find the mean value of the ratio E_1/E_2.

COMPARISON OF ELECTROMOTIVE FORCES BY THE CONSTANT RESISTANCE METHOD.

Apparatus.—Galvanometer, resistance, key, two cells whose E.M.F.'s are to be compared.

Method.—The E.M.F.'s may be compared by means of the currents they send through the galvanometer when they are separately in series with it and a resistance, this resistance being large compared with the internal resistance of the cells.

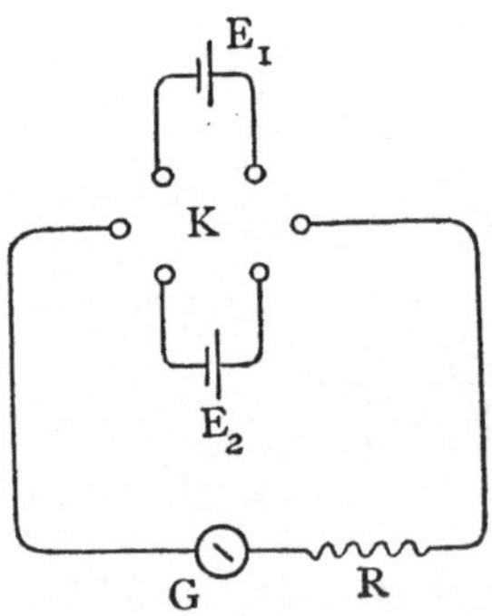

FIG. 55.

Connect the cells E_1 and E_2 as in the figure in series with a resistance R and a galvanometer G, so that by means of the key K each cell can be separately put in the circuit.

When E_1 is in the circuit, adjust R, so that a suitable deflection d_1 is got on the galvanometer (it may be necessary to shunt the galvanometer): with all the conditions the same, throw over the key, cutting out E_1 and putting in E_2, and observe the deflection d_2. Repeat for various values of R.

The current in the first case

$$C_1 = \frac{E_1}{R+G+B_1} = kd_1 \text{ or } k \tan d_1.$$

In the second case

$$C_2 = \frac{E_2}{R+G+B_2} = kd_2.$$

Thus, neglecting the battery resistances B_1 and B_2 in comparison with R+G,

$$\frac{E_1}{E_2} = \frac{d_1}{d_2}, \text{ or } \frac{\tan d_1}{\tan d_2}.$$

COMPARISON OF ELECTROMOTIVE FORCES BY A POTENTIOMETER.

Apparatus.—Galvanometer, potentiometer (wire of metre bridge), two cells whose E.M.F.'s are to be compared, and a source of constant greater E.M.F.

Method.—(As this is a null or compensation method of comparing E.M.F.'s, it is eminently suited for cells which are liable to polarise, *e.g.*, the Leclanché, since no current is flowing when balance is attained.)

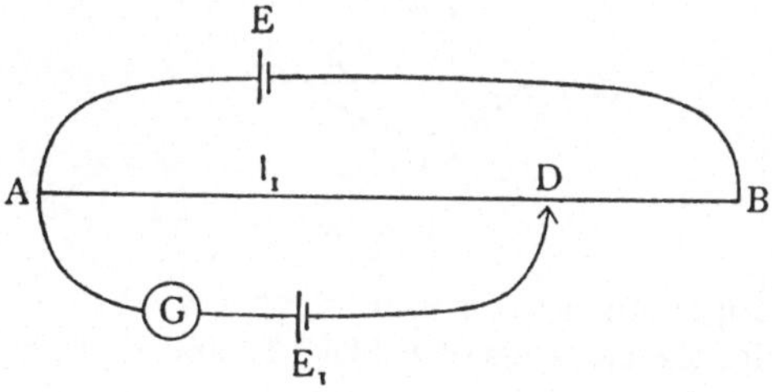

FIG. 56.

Connect as in the diagram. Let E_1 and E_2 be the E.M.F.'s of the two cells to be compared and E (greater than E_1 or E_2) the E.M.F. of the auxiliary cell. The cell (E) is connected to the ends A and B of the wire of the bridge, and (E_1) is connected through the galvanometer G to the one end of the bridge wire, the *positive* or *negative* terminals of both cells being connected to A. The other terminal of (E_1) is connected to the slide of the bridge wire, and this slide is moved along the wire to D till no deflection is observed in G. When this is the case, then the potential difference between the points A and D is equal to the E.M.F. E_1. If the wire is uniform, Ohm's Law gives

$$E_1 = r_1 C = kl_1 C,$$

where C is the current through the wire, l_1 is the length AD, and r_1 or kl_1 is its resistance.

Repeat, using the cell (E_2) instead of (E_1), and find a position D′ so that no current flows through the galvanometer G, and let AD′ $= l_2$, then

$$E_2 = kl_2 C; \qquad \text{hence } \frac{E_1}{E_2} = \frac{l_1}{l_2}.$$

It is important that the current in the wire should remain constant during the two observations, hence the necessity for a cell (E) of constant E.M.F.

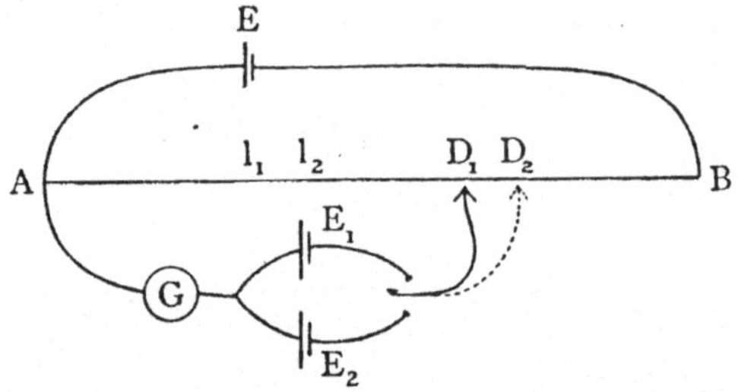

FIG. 57.

An arrangement shown in Fig. 57 may be used, in which a key enables the two cells of E.M.F., E_1 and E_2, to be alternately connected to D.

ELECTROCHEMICAL EQUIVALENT OF A METAL.

Apparatus.—Copper voltameter, battery, resistance, tangent galvanometer, key.

Method.—Arrange the apparatus as in the diagram, the battery B, voltameter V, galvanometer G, resistance R, and key K being in series.

The current should be made to flow in such a direction as to produce deposition of copper on the copper vessel. This, accordingly, should be the negative pole of the electrolytic cell.

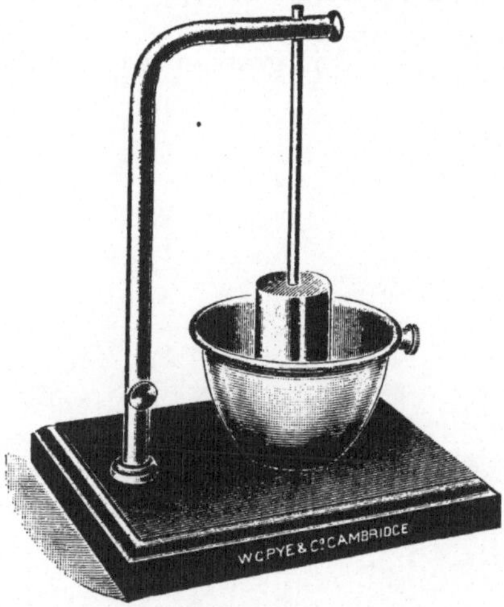

FIG. 58.

Weigh the copper vessel, fill it with copper sulphate solution, and insert the other copper electrode. Then pass the current for about half an hour, noting the time during which the current has been flowing. Observe at intervals the reading on the

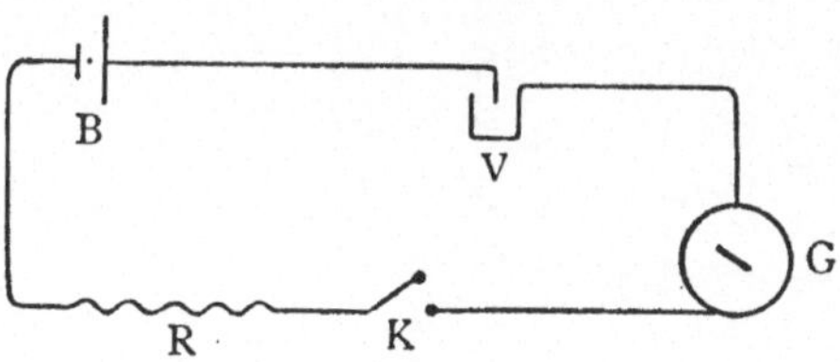

FIG. 59.

tangent galvanometer, and, if necessary, adjust the resistance R, to keep the current constant. Stop the current and weigh the copper vessel after it has been washed and dried. The constant of the tangent galvanometer being known, the current in amperes (C) can be obtained.

If W_1 and W_2 grams are the respective weights of the copper vessel before and after the electrolysis, t the time in seconds during which the current of C amperes has been running, then the electrochemical equivalent of copper is

$$\frac{W_2 - W_1}{Ct} \text{ in grams per coulomb.}$$

COMPARISON OF ILLUMINATING POWERS BY THE SHADOW PHOTOMETER.

Apparatus.—Two sources of light (say candle and lamp), vertical rod, vertical white screen.

Method.—Place the opaque rod R in front of the screen at a distance of say 9 inches or 1 foot. Arrange the sources of light S_1 and S_2, so that two shadows of the rod are cast near together upon the screen, and vary the distances of the sources from the screen until the shadows D_1 and D_2 appear as nearly

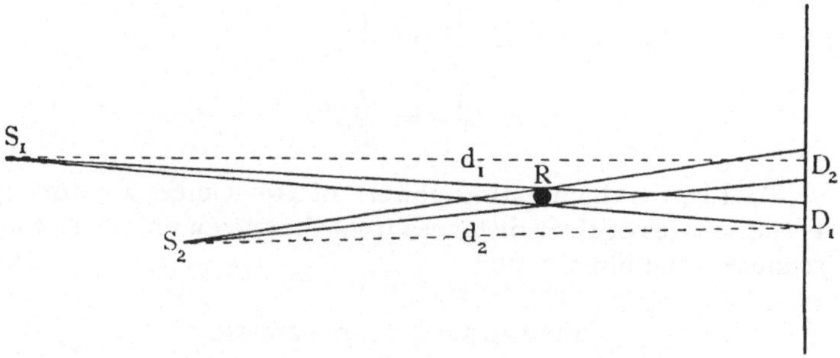

FIG. 60.

as possible of the same density. (Owing to the different colours of the lights from different sources, there is frequently some difficulty in deciding when the illuminations of the two shadows are equally intense. The difference in tint is, however, much less disturbing when the eyes are partially closed.)

The shadow D_1 is illuminated by the source S_2, and shadow D_2 is illuminated by the source S_1. If I_1 and I_2 be the illuminating powers of the sources S_1 and S_2, the intensities of illumination at D_2 and D_1 are

$$\frac{I_1}{d_1^2} \text{ and } \frac{I_2}{d_2^2}$$

where $S_1D_2 = d_1$ and $S_2D_1 = d_2$.

But the intensities have been made equal,

$$i.e., \frac{I_1}{d_1^2} = \frac{I_2}{d_2^2},$$

$$\therefore \frac{I_1}{I_2} = \frac{d_1^2}{d_2^2}.$$

That is, the illuminating powers of two sources are directly as the squares of their distances from the screen on which they produce equal illumination.

GREASE-SPOT PHOTOMETER.

Apparatus.—Grease-spot photometer, lights to be compared (*e.g.*, candles, lamp, etc.), measuring scale.

Method.—The two sources of light to be compared are placed on opposite sides of the photometer on a level with the grease-spot, and their distances from the photometer are so adjusted that the grease-spot and the rest of the paper, when viewed from a short distance, seem equally bright.

This occurs when the light going through the grease-spot in one direction is balanced by that coming through in the opposite direction. But if a surface is illuminated by two sources of light of illuminating powers I_1 and I_2 when these are at distances d_1 and d_2 from the surface, then the intensities of illumination are given by

$$\frac{I_1}{d_1^2} \text{ and } \frac{I_2}{d_2^2}.$$

These have been made equal,

$$\therefore \frac{I_1}{I_2} = \frac{d_1^2}{d_2^2};$$

hence by measuring d_1 and d_2 we can find the ratio of I_1 to I_2.

LAWS OF REFLECTION (Pin Experiments).

To show that the Angle of Reflection is equal to the Angle of Incidence for a Plane Mirror.

Apparatus.—Plane mirror mounted on stand so that it is perpendicular to the base, drawing board, sheet of white paper, pins.

Method.—A straight line AB is drawn on the sheet of paper, and the mirror is placed with its back reflecting surface over this line. In front of it place two pins through points P_1 and P_2 on the paper; then insert two pins in positions Q_1 and Q_2, such that on looking along Q_1Q_2 the images of P_1 and P_2 appear on the same straight line as Q_1Q_2. Join P_1P_2 and Q_1Q_2, and let these intersect in M. [If the mirror is thick, this will be appreciably in front of AB owing to refraction in the mirror itself.] Draw MN normal to AB at M, and show that the angles P_1MN and Q_1MN are equal. This may be done by means of a protractor, or by making a construction which measures some trigonometrical function of the angles.

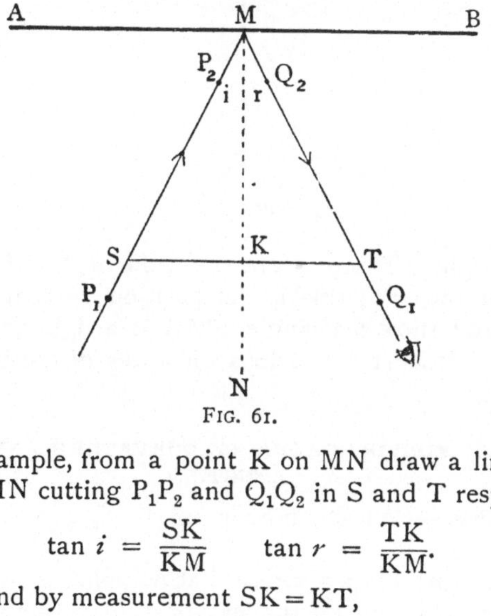

FIG. 61.

[For example, from a point K on MN draw a line perpendicular to MN cutting P_1P_2 and Q_1Q_2 in S and T respectively.

$$\tan i = \frac{SK}{KM} \qquad \tan r = \frac{TK}{KM}.$$

It is found by measurement SK = KT,

$$\therefore \quad i = r.]$$

Repeat for various angles of incidence.

The Image of a Point is formed on the Normal to a Plane Mirror, and is as far behind the Mirror as the Point is in Front.

Apparatus.—Plane mirror mounted on stand, drawing board, paper, pins.

Method.—Place a pin in a vertical position through a point N in front of the mirror, and behind the mirror insert another and longer vertical pin through a point N′, so that on looking at the mirror, this pin appears to be a continuation of the image

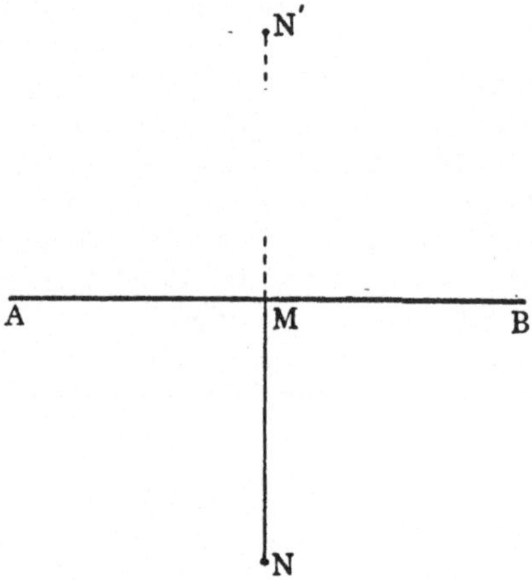

FIG. 62.

of the first pin. There is one position of N′ which suits all positions of the eye; this is the position N′ that is required. Join NN′ and show that angle NMA is a right angle and that MN = MN′. Repeat for various distances of the pin from the mirror.

CONCAVE MIRROR: RADIUS OF CURVATURE AND FOCAL LENGTH.

Apparatus.—Concave mirror on stand, two large pins on stands.

Method.—(*a*) Place a pin so that its point is approximately on the principal axis of the mirror, and with the eye much farther from the mirror, observe the position of the image of the pin. Move the pin backward or forward along the principal axis until the object A and image A′ coincide, that is, till the image of the pin point appears to coincide with the point of the object. This coincidence is obtained by the parallax method. (As the eye is moved across the principal axis, the object and

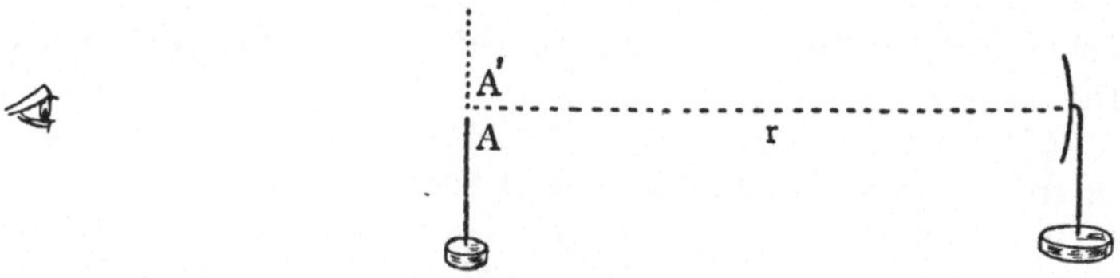

FIG. 63.

image appear to move together across the face of the mirror without any visible separation.) When this is attained, the point must be at the centre of curvature of the mirror, rays proceeding from this point falling normally on the mirror and being reflected back along their original path.

The distance from the pin point to the mirror is the radius of curvature. The focal length is half of this.

(Repeat the experiment a number of times and take the mean result.)

(*b*) Instead of making the object and image coincide, move a second pin B along the principal axis, till the point of B coincides with the image A′ of the point of A. This again is

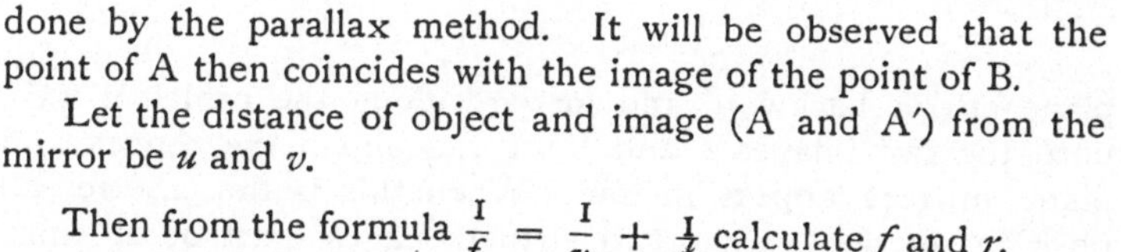

FIG. 64.

done by the parallax method. It will be observed that the point of A then coincides with the image of the point of B.

Let the distance of object and image (A and A′) from the mirror be u and v.

Then from the formula $\frac{1}{f} = \frac{1}{v} + \frac{1}{u}$ calculate f and r.

When the distance from mirror to object is less than the focal length f, the image is virtual and is behind the mirror.

The position of the virtual image may, however, be obtained by placing an upright pin B so that a portion of B as seen above or below the mirror appears in line with the image of A as seen in the mirror. In using the formula $\frac{1}{f} = \frac{1}{v} + \frac{1}{u}$, the value of v is in this case negative, being on the opposite side to the object.

(*c*) Use a luminous body as the object (lamp or candle flame, or cross wire illuminated by one of these). When the object is farther from the mirror than the focus, the image is real, and may be received on a small white screen, which should be moved until the image is clearly defined. The source and screen must, of course, be placed eccentrically, the two being on opposite sides of the principal axis.

Use the formula $\frac{1}{f} = \frac{1}{v} + \frac{1}{u}$ as above for various positions of the object.

(*d*) Determine r by the spherometer, and calculate f

CONVEX MIRROR: RADIUS OF CURVATURE AND FOCAL LENGTH.

Apparatus.—Convex mirror, plane mirror, large pin on stand.

Method.—Place the pin P in front of and at a fair distance from the convex mirror M, and with the eye farther from the mirror and near its principal axis, arrange the plane mirror M, to cover about half the face of the convex mirror. Move the

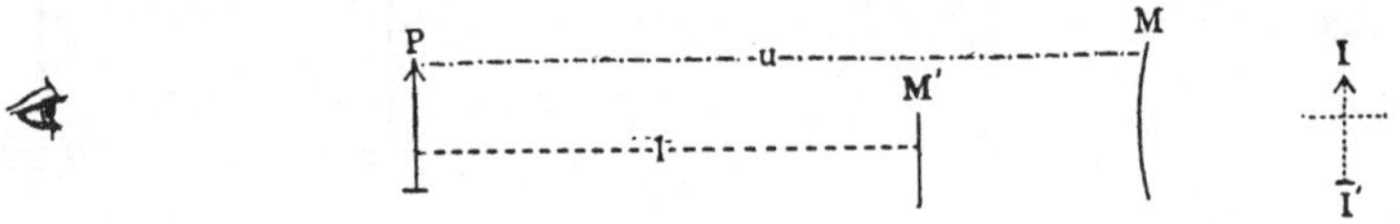

FIG. 65.

plane mirror backward and forward along the principal axis, until the two images I and I′ of the pin in the convex and plane mirrors appear in line. When this is the case for all positions of the eye, the two virtual images must be at equal distances from the convex mirror. But we know the position of the image in a plane mirror, hence we have determined the position of the image in the convex mirror.

Thus if u be the distance of the pin P from the convex mirror, and l the distance between the pin and the plane mirror, the distance of the images from the plane mirror is l. Hence the distance of the image from the convex mirror is $2l-u$. Call this v.

Then the formula for mirrors $\frac{1}{f} = \frac{1}{v} + \frac{1}{u}$ enables us to calculate f, the focal length of the convex mirror. It is negative in sign. The radius of curvature $r=2f$.

LAWS OF REFRACTION.

The Ratio of the Sine of the Angle of Incidence to the Sine of the Angle of Refraction is Constant (the Refractive Index).

Apparatus.—Plate of glass with parallel faces, drawing board, paper, pins.

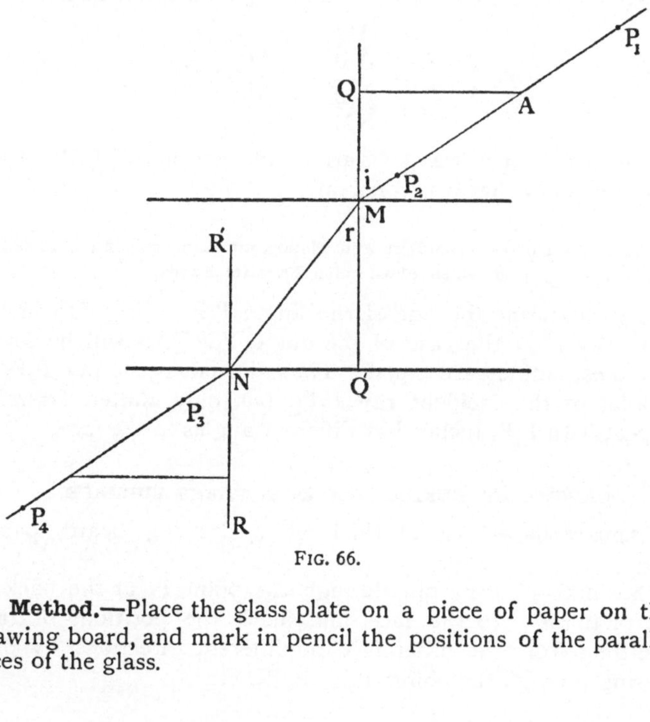

FIG. 66.

Method.—Place the glass plate on a piece of paper on the drawing board, and mark in pencil the positions of the parallel faces of the glass.

Insert two pins in positions P_1 and P_2 on one side of the plate, so that P_1P_2 makes a considerable angle with the normal to the face of the plate. Place two other pins in positions P_3 and P_4 on the other side of the plate, so that on looking along the line P_4P_3, the pins P_4 and P_3 appear in the same straight line as the images of P_2 and P_1 when viewed through the glass plate.

Let the lines P_1P_2 and P_4P_3 intersect the respective faces in M and N. Join MN. The ray P_1P_2 thus enters the glass at M, is refracted in the direction MN, and emerges as the ray P_3P_4.

At M and N draw normals QMQ′ and RNR′ to the faces of the plate. Then P_1MQ is the angle of incidence (air to glass) and NMQ′ the angle of refraction;

$$\therefore \quad \frac{\sin i}{\sin r} = \mu \text{ (air to glass)} = \frac{\sin P_1MQ}{\sin NMQ'}.$$

If A is such that MA = MN and AQ is perpendicular to QQ′,

$$\text{then} \qquad \mu = \frac{\sin i}{\sin r} = \frac{\frac{AQ}{AM}}{\frac{NQ'}{NM}} = \frac{AQ}{NQ'}.$$

Determine this ratio for various inclinations of P_1P_2 to the face, and show that it is constant.

Parallelism of the Incident and Emergent Rays passing through a Glass Plate with Parallel Faces.

By measuring the sine of the angle P_4NR (Fig. 66) in the same way as in the case of the angle P_1MQ, it will be found that these angles are equal. Thus the emergent ray P_3P_4 is parallel to the incident ray P_1P_2, but only shifted laterally. Repeat with P_1P_2 inclined at different angles to the face.

CAUSTIC BY REFRACTION AT A PLANE SURFACE.

Apparatus.—Plate of thick glass, drawing board, paper, pins.

Method.—Place a pin through the point P_1 at the back of the plate close to the face, and mark the positions of rays emerging from the other face in different directions by pins passing through the points P_2P_3, $P_2'P_3'$, etc.

Join P_3P_2, $P_3'P_2'$, etc., and produce these lines. The curve which touches all these rays produced is the caustic, its cusp being at the image of P_1 when viewed normally through the glass.

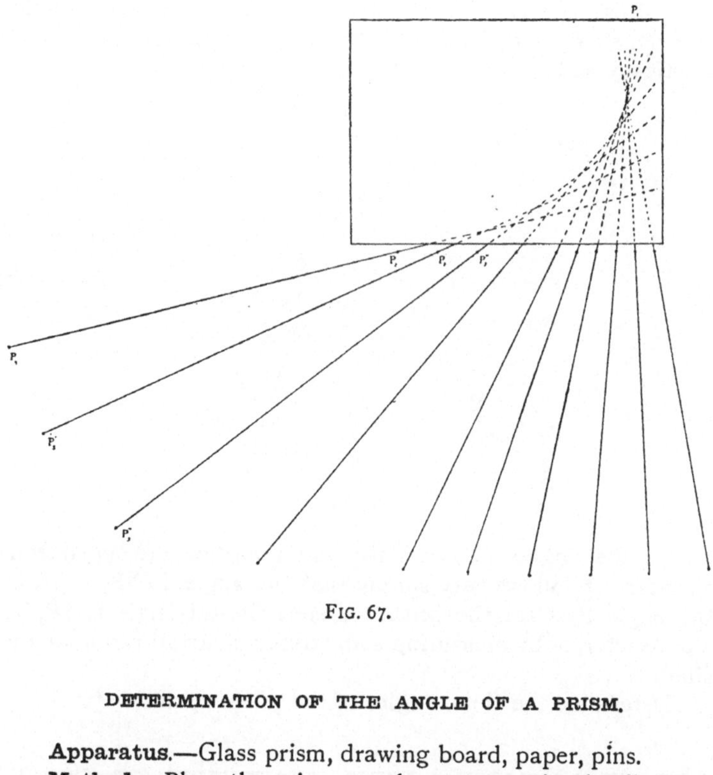

FIG. 67.

DETERMINATION OF THE ANGLE OF A PRISM.

Apparatus.—Glass prism, drawing board, paper, pins.

Method.—Place the prism on the paper as in the diagram, and fix a pin through a point P in a position not too near the edge of the prism at which the angle is to be measured.

Place the eye near the paper in a position such as F, so as to see the image of the pin reflected in the face AB of the prism, and move the eye until the image is seen in a portion of the face AB as near the A end as possible. Fix two pins through points P_1 and P_2 on the board, so that P_2, P_1, and the image of P are in the same straight line.

Similarly, on the other side of the face AC, place two pins through points P_4 and P_3, so that P_4, P_3, and the image of P in the face AC appear in the same straight line, reflection again being from the end of face AC as near to A as possible. Join P_2 to P_1 and P_4 to P_3. By the construction they meet practically at A.

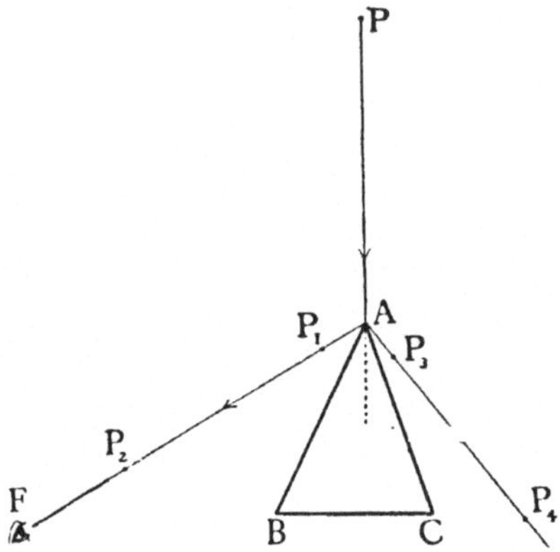

FIG. 68.

As the angles of incidence and reflection are equal from each face, it follows very simply that the angle P_2AP_4 is twice the angle BAC of the prism. Determine the angle P_2AP_4 by a protractor, or by measuring some trigonometrical ratio, say the sine.

Determine the three angles of the prism in this way.

REFRACTION THROUGH A PRISM: MEASUREMENT OF MINIMUM DEVIATION AND REFRACTIVE INDEX.

Apparatus.—Glass prism, drawing board, paper, pins.

Method.—Place the prism on the drawing board so that the trace of the prism is as represented in the figure. Place two pins passing through points P_1 and P_2 in positions as represented. P_1P_2 is the direction of the incident ray studied.

Looking through the other face of the prism, observe the direction of the emergent ray by placing the eye on the line of the images of P_1 and P_2, that is, on the line P_3P_4 of the emergent ray. Slowly rotate the prism about a vertical axis, and follow the direction of this emergent ray. It will be seen to depend on the position of the prism. As this is rotated in the proper direction, the line P_3P_4 will be seen to turn nearer to the line P_1P_2 produced till the angle between P_1P_2 produced and P_3P_4 reaches a minimum, *the minimum deviation;* it then increases again. Finally, set the prism in this position of minimum deviation, and place two pins through points P_3 and P_4 in the paper, so that when looking along P_4P_3 through the prism, the points P_1P_2 appear in the same line. (The pins if vertical appear in the same plane.)

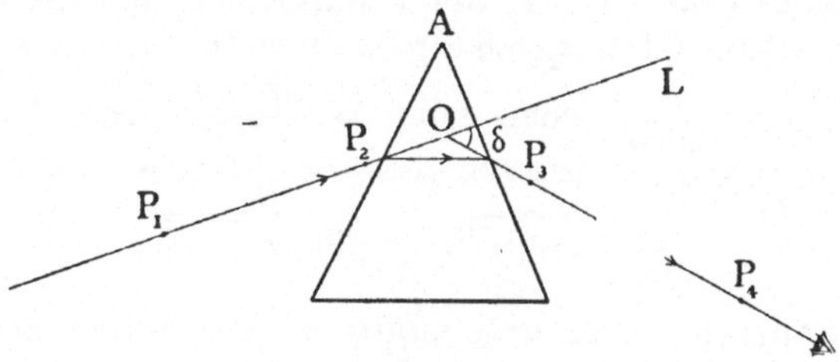

FIG. 69.

Produce P_1P_2 and P_3P_4 to intersect in O. Angle P_4OL is the minimum angle of deviation. It will be observed that the angle of incidence is equal to the angle of emergence, that is, the prism is symmetrically placed with regard to the rays P_1P_2 and P_4P_3.

Calling the minimum deviation δ—(the different constituents of white light suffer somewhat different deviations)—and the angle of the prism a, the refractive index μ of the glass is given by the formula

$$\mu = \frac{\sin \frac{1}{2}(a+\delta)}{\sin \frac{1}{2}a}.$$

REFRACTIVE INDEX OF THE MATERIAL OF A PRISM BY TOTAL REFLECTION.

Apparatus.—Prism, drawing board, paper.

Method.—Draw two parallel lines a few centimetres apart on the paper, and place the paper on the drawing board. Place the prism with its edge along the line AB, and covering the other line CD as in the figure.

Look through the face of the prism at the line CD, and move the eye downwards till CD just disappears. Note where the

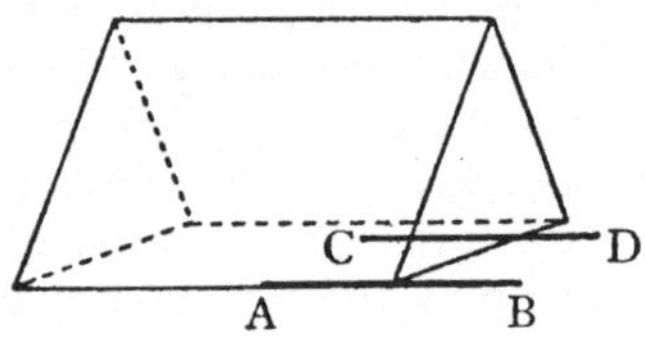

FIG. 70.

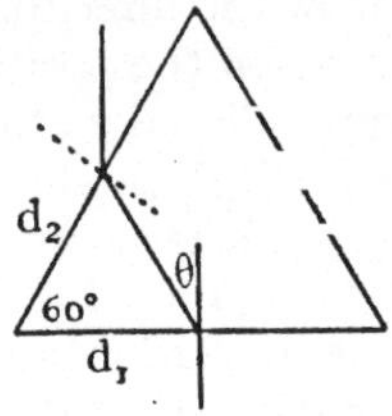

FIG. 71.

vanishing ray emerges from the face of the prism, and measure the distance from AB. If this distance is d_2 and the distance between AB and CD is d_1 (see Fig. 71), then the refractive index of the glass $\mu = \frac{1}{\sin\theta}$, where $\tan\theta = \frac{2d_1 - d_2}{\sqrt{3}d_2}$ (for a 60° prism)

$$\text{for} \quad \frac{d_1}{\sin(30^\circ + \theta)} = \frac{d_2}{\sin(90^\circ - \theta)}.$$

REFRACTIVE INDEX OF A LIQUID BY TOTAL REFLECTION.

Apparatus.—Glass cell consisting of two parallel glass plates with a thin layer of air between, cubical vessel to contain liquid, drawing board, paper, pins.

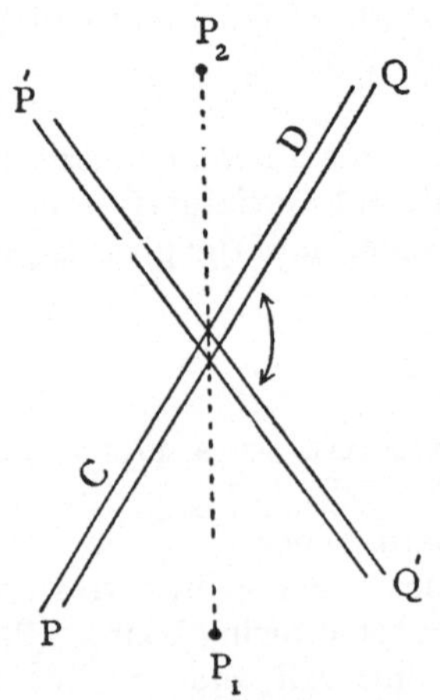

FIG. 72.

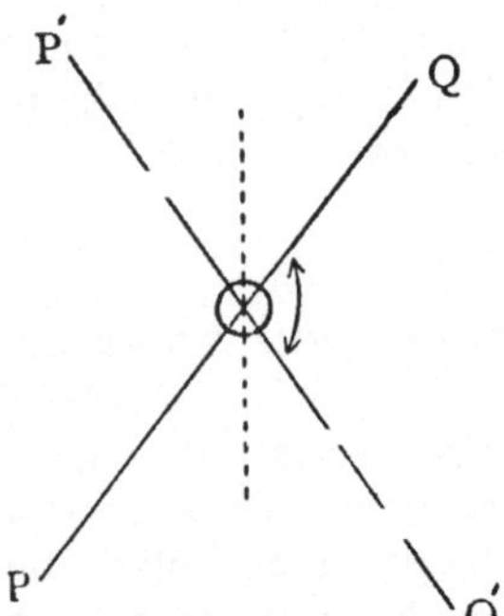

FIG. 73.

Method.—Place the apparatus in the glass vessel on the drawing board with two pins P_1 and P_2 as in the figure, and turn the apparatus CD about a vertical axis till the pin P_2, as viewed from P_1 through the glass cell, disappears from view. Mark the position PQ of the cell CD where this occurs, and then turn the cell round till this again occurs at P′Q′.

Remove the apparatus and join PQ, P′Q′ intersecting in O. The refractive for the liquid

$$\mu = \frac{1}{\text{sin of critical angle}} = \frac{1}{\sin \frac{QOQ'}{2}}.$$

REFRACTIVE INDEX OF A LIQUID BY A CONCAVE MIRROR.

Apparatus.—Concave mirror, pin, transparent liquid.

Method.—Place the mirror on a low table with its face

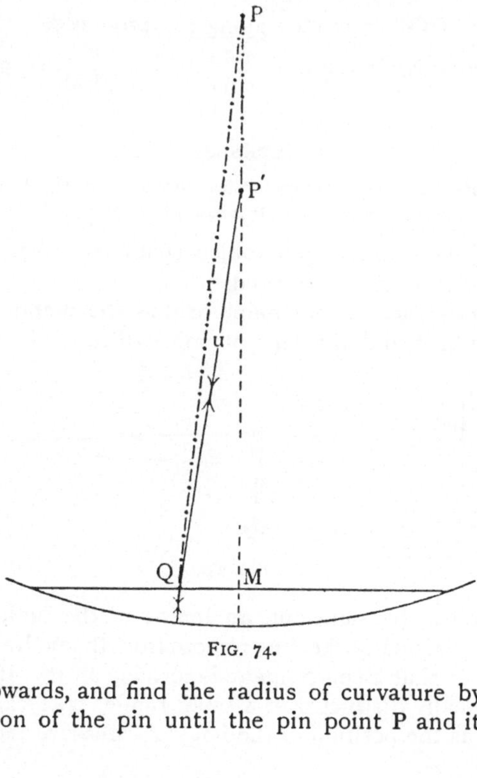

FIG. 74.

turned upwards, and find the radius of curvature by adjusting the position of the pin until the pin point P and its image in the concave mirror, as seen from above the pin, coincide (see experiment p. 88). The distance of the pin from the mirror is r, the radius of curvature of the mirror.

Pour a small quantity of the liquid on the face of the mirror. The liquid forms a plano-convex lens.

Adjust the position of the pin again until the image of the point coincides with the new position (P′) of the pin point. Let the distance from the mirror now be u. In this case, as before, rays from the point retrace their path—hence after passing through the liquid lens, they fall normally on the face of the mirror. If the mirror were absent, they would pass normally through the second (curved) surface of the liquid lens. Therefore rays from an object distant u from the liquid surface proceed after refraction as though they had radiated from a distance r.

Thus ray P′Q is refracted along PQ produced.

But the refractive index $\mu = \dfrac{\sin i}{\sin r} = \dfrac{\sin QP'M}{\sin QPM} = \dfrac{QP}{QP'} = \dfrac{r}{u}.$

LENSES.

Relation between the Distance of an Object and its Image from a Thin Convex Lens; and Determination of Focal Length.

Apparatus.—Convex lens, object (pinhole in sheet of metal), gas flame, screen, measuring rod.

Method.—Place the gas flame behind the metal sheet with the pinhole in it and the lens at some distance in front of it.

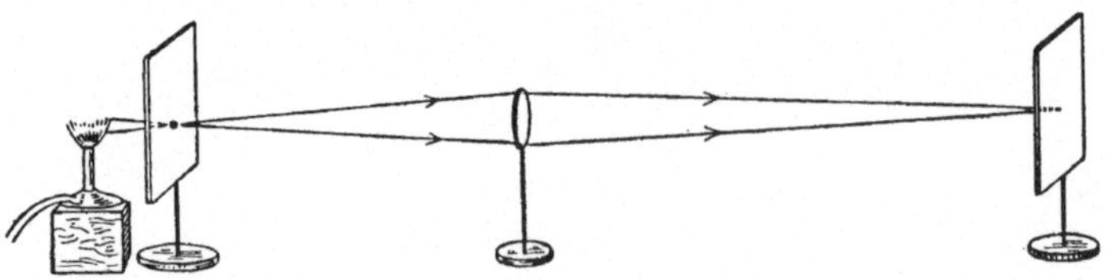

FIG. 75.

If this distance be sufficient, an image of the pinhole will be formed on the side of the lens remote from the pinhole. Adjust the screen so that a sharp image is formed on it. If the image appears equally distinct over a small range, take the centre of the range as the position of the image. Measure the distances between the lens and the object (U) and the lens and the image (V). Do this for some half-dozen positions.

The relation between u and v, the distances of object and image from the lens when account is taken of direction, is $\frac{1}{v} - \frac{1}{u} = \frac{1}{f}$ where f the focal length of the convex lens is negative, and v is negative if a real image is formed. (*N.B.*—Regard distances from the lens in direction opposite to that of incident light as positive.)

Hence test the relation

$$\frac{1}{U} + \frac{1}{V} = \frac{1}{F},$$

where U, V, F express distances but do not take account of the direction in which they are measured, that is, $U = u$, $V = -v$, $F = -f$.

The values of $\frac{1}{U} + \frac{1}{V}$ should be constant.

This can also be shown graphically by plotting values of U along the x axis and values of V along the y axis, and joining corresponding points on the x and y axes. These will all pass approximately through one point P such that $Op = pP$ (where p is the foot of the perpendicular from P on the x axis) = numerical value F of the focal length of the lens; for the equation to the line cutting off intercepts U and V from the axes is $\frac{x}{U} + \frac{y}{V} = 1$. But for all pairs of values of U and V this line passes through the point (F, F);

$$\therefore \quad \frac{F}{U} + \frac{F}{V} = 1 \qquad \textit{i.e.,} \quad \frac{1}{U} + \frac{1}{V} = \frac{1}{F}.$$

Focal Length of Convex Lens (Pin Method).

Apparatus.—Convex lens and holder, two pins on vertical stands, metre scale.

Method.—Place one pin P_1 a little distance from one face of the lens, and with the eye on the other side of the lens, and at a distance from this observe the image P_1' of this pin. If the pin is farther away from the lens than the focal length, the image will be real, inverted, and on the opposite side of the lens, *i.e.*, on the eye side. Place the second pin P_2 on this side, and move it backward and forward along the principal axis of the lens until it appears by the side of the image P_1' of the first pin. Adjust so that the points of P_1' and P_2 practically coincide, and there is no relative motion of the two when the eye is moved from side to side. Pin P_2 thus gives the position of the image P_1' of P_1.

FIG. 76.

Measure distances of P_1 and P_2 from the lens, then calling these u and v we have, paying regard to signs

$$\frac{1}{v} - \frac{1}{u} = \frac{1}{f}.$$

Calculate f from this equation.

Repeat the experiment for various positions of the object.

Focal Length of Concave Lens (Pin Method).

Apparatus.—Concave lens on stand, two large pins on stands.

Method.—The method used for a concave lens is similar in principle to that for a convex lens. The image, however, is virtual and on the same side of the lens as the object. As in the experiment with a convex lens the object (pin) and eye are on opposite sides of the lens. To find the position of the virtual image, the second pin is placed on the same side of the lens as the first, but nearer the lens. It must, however, be viewed either above or below the lens, and the position altered until the image of the first pin as viewed through the lens appears in line with the second pin as viewed above or below the lens. By trial a position will be found in which this is the case even when the eye is moved from side to side. The second pin then gives the position of the image of the first pin.

Then if u and v be the distances of the first pin (object), and the second pin (giving position of image) from the lens,

$$\frac{1}{v} - \frac{1}{u} = \frac{1}{f}.$$

Obtain f, and repeat the experiment for various values of u and v. (In the case of a concave lens f is positive.)

Focal Length of a Concave Lens by Combination with a Convex Lens.

Apparatus.—As in the determination of focal length of convex lens, with the addition of a concave lens.

Method.—Owing to the fact that a concave (diverging) lens gives virtual images, the method of obtaining images on a screen, used in the case of the convex lens, cannot be applied to a concave lens alone.

By combining the concave lens with a stronger convex one, a combination may be got which is a convergent lens.

Then the relation between the focal lengths of the convex f_1, concave f_2, and combination F, when the lenses are thin and in contact, is

$$\frac{1}{F} = \frac{1}{f_1} + \frac{1}{f_2}$$

$$\frac{1}{f_2} = \frac{1}{F} - \frac{1}{f_1}$$

$$\therefore \quad f_2 = \frac{f_1 F}{f_1 - F}.$$

(*N.B.*—Convex or converging lenses have negative focal lengths. Concave or diverging lenses have positive focal lengths.)

Hence by determining the focal lengths (F and f_1) of the two converging lenses, the value of f_2 can be calculated.

MAGNIFYING POWER OF SIMPLE MICROSCOPE.

Apparatus.—Convex lens of short focal length, two clamps on stands adjustable vertically—one for lens, and the other for one scale, two accurately divided scales.

Method.—Place one scale (*a*) on the table, and clamp the other one (*b*) almost vertically above it. Looking through the lens with one eye in a position quite close to the lens, adjust the height of the scale *b* and the lens until the image of *b* as seen through the lens, and the actual scale *a* as seen directly with the other eye, appear superposed. This can be obtained by the method of parallax—a side movement of the eyes produces no relative motion of the image of *b* and *a*.

When the two scales are thus superposed, the number of divisions on *a* enclosed between two corresponding divisions of the image of *b* may be read directly. This gives the magnification.

VII

SOUND.

DETERMINATION OF FREQUENCY OF A NOTE BY THE SIREN.

Apparatus.—Tuning-fork, siren, foot-bellows, stop-watch.

Method.—Strike the tuning-fork, the frequency of vibration of which is to be determined, and while holding it near the ear, regulate the supply of air from the foot-bellows to the siren until it emits approximately the same note. The regulation should be performed by the foot.

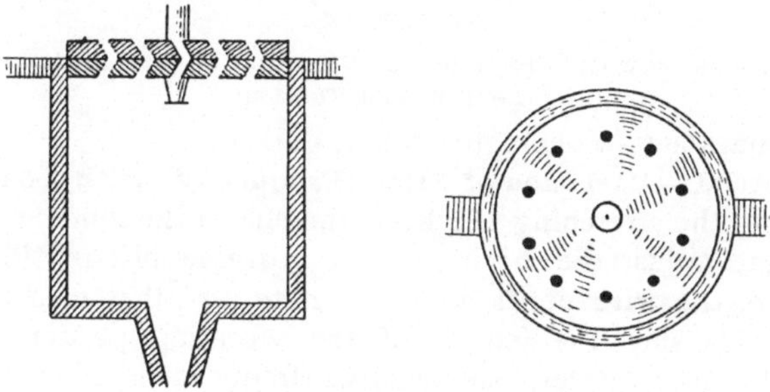

FIG. 77.

Observe the readings on the indicator of the siren, and when it reaches a definite number of thousands, say 7000, start the stop-watch. By repeatedly striking the fork and working the bellows continuously, the note of the siren should be maintained as near in pitch to that of the fork as possible for a period of about a minute, say. The stop-watch should then be stopped when the siren indicator reads another whole number of thousands.

Thus N revolutions of the siren disc given by the indicator have been made in the time t given by the watch. If there are n holes in the siren disc, n puffs of air pass through it in each revolution. Thus the frequency of the puffs, which is the same as that of vibration of the fork, is Nn/t.

(Some difficulty will be found in accurately maintaining the pitch of the siren by the bellows. An attempt should therefore be made to make slight variations of pitch to compensate, a slight accidental flattening of the pitch being balanced by a slight subsequent voluntary sharpening.)

SONOMETER.

Relations between Frequency of Vibration of Stretched String, Length, and Tension.

Apparatus.—Sonometer, tuning-fork.

Method.—Variation of Pitch (Frequency) with Length.—Keeping the stretching weight on the end of the sonometer wire constant, adjust the position of a movable bridge, till when plucked, the wire emits a certain note, say that of a tuning-fork. Measure the length of the vibrating portion. Then tune to other notes, the relative frequencies of which are known, by simply moving the bridge, care being taken that the string runs freely over the bridge, thus ensuring equal tension on the two sides of the bridge. If only one tuning-fork is available, tune the string to the fork and afterwards to the major third, perfect fifth, and octave, and so on above the pitch of the fork. Thus, if n is the frequency of the fork, the frequencies to which the string is tuned are n, $\frac{5}{4}n$, $\frac{3}{2}n$, etc. Tabulate with the corresponding length of vibrating string.

	Frequency.	Length.	Product of Length and Frequency.
doh	n		
me	$\frac{5}{4}n$		
soh	$\frac{3}{2}n$		
doh′	$2n$		
me′	$\frac{5}{2}n$		
soh′	$3n$		

It will be found that the length is inversely proportional to the frequency, or that frequency × length of string is approximately constant.

Variation of Pitch (Frequency) with Tension.—Tune the string successively to notes of known frequency, or known relative frequency (n, $\frac{5}{4}n$, $\frac{3}{2}n$, etc., say), by varying the stretching weight, while the length is kept constant. It will be found that the frequency varies as the square root of the stretching force,

$$i.e., \quad n \propto \sqrt{T}$$

$$\text{or} \quad \frac{n^2}{T} \text{ is constant.}$$

Variation of Length with Tension for a Constant Pitch (Frequency). — Make a series of observations with various weights suspended from the end of the string, keeping the pitch constant by moving the bridge, and thereby varying the length of the vibrating string. It will be found that the tension is proportional to the square of the length of string when the pitch is kept constant, *i.e.*, $T \propto l^2$ or T/l^2 is a constant. Tabulate results.

Theory shows that the frequency

$$n = \frac{1}{2l}\sqrt{\frac{T}{m}},$$

where m is the mass per unit length.

RESONANCE TUBE.

Apparatus.—Tuning-fork, long wide glass tube (say inch wide) connected by flexible tubing to water bottle, metre scale.

Method.—While holding a vibrating fork above the open end of the resonance tube, gradually lower the level of the water surface in the tube by lowering the reservoir. It will be observed that in certain positions of the water surface, the sound is strongly reinforced by resonance. Adjust rubber bands round the tube to these positions, and let the distances from the top of the tube be l_1, l_2, l_3, etc. The distance between consecutive positions of resonance is half a wave-length of the note emitted by the tuning-fork; while the distance from the top end to the first resonance position is a little less than a quarter of a wave-length. Measuring from a point at a distance $\frac{3}{5}r$ outside the open end of the tube (r being the internal radius of the tube), the distances of the positions of resonance are $\lambda/4$, $3\lambda/4$, $5\lambda/4$, etc. Determine the wave-length of the note in air from measurement of the first position of resonance, *i.e.*, from

$$(l_1 + \tfrac{3}{5}r) = \frac{\lambda}{4}.$$

Also find λ from the equations $l_2 - l_1 = l_3 - l_2 = \frac{\lambda}{2}$.

Knowing the frequency of the fork (n), and the wave-length of the sound waves in air (λ), the velocity of sound through air (V) is given by $V = n\lambda$. Thus velocity is, of course, that through air contained in the tube at the temperature of the air in the particular experiment.

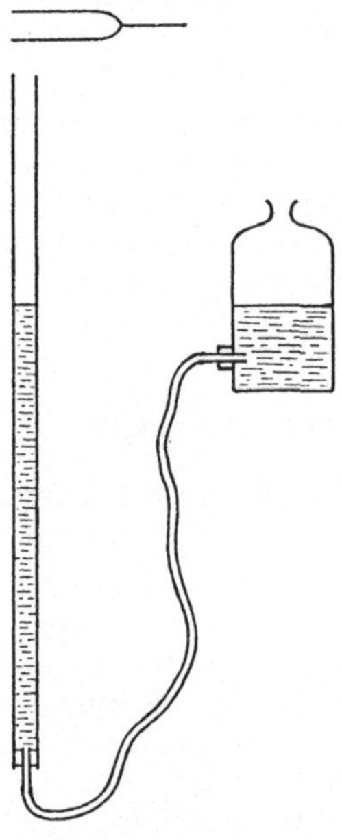

FIG. 78.

Or knowing the velocity of sound in air at the particular temperature, and determining the wave-length by measurement as above, the frequency of the particular tuning-fork may be obtained.

The end correction (x), assumed above to be $\frac{3}{5}r$, for the tube may be obtained by this experiment, and the application of the formula $(l_1 + x) = \frac{1}{2}(l_2 - l_1)$.

KUNDT'S TUBE.

Apparatus.—Glass tube a metre or more in length and an inch or so internal diameter; narrower glass, brass and steel tubes, clamp, lycopodium powder, light cardboard disc to fit loosely in wider tube.

Method.—Fit the light disc on the end of the narrower glass tube and clamp this tightly at the middle, so that the end with disc is just inside the larger glass tube, which is coaxial with the first. The disc must be perfectly free to move inside the larger tube. Lycopodium powder, or other light powder, should be spread as uniformly as possible along the bottom of the larger tube.

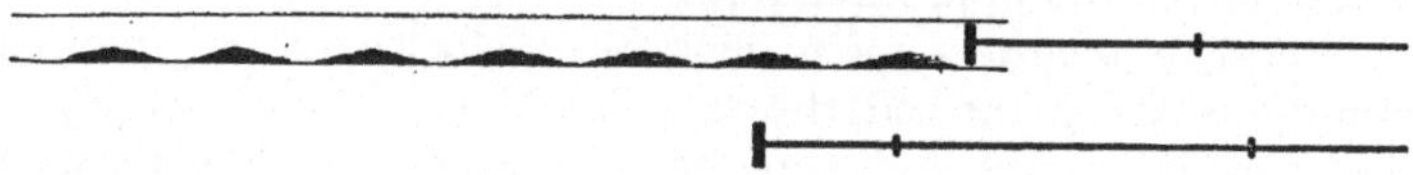

FIG. 79.

Stroke the end of the narrower tube with a cloth moistened with methylated spirits until it is set in longitudinal vibratory motion and emits a high note. Move the wider tube backward and forward a little over the vibrating disc, until the air in this tube vibrates in resonance with the vibrating rod. This is shown by the most violent motion of the powder inside the tube; it becomes heaped up in positions regularly distributed along the tube. These are the nodes—places of minimum disturbance in the air—the powder being driven from the antinodes where the disturbance is greatest, to the nodes where the air is undisturbed. The positions are most clearly shown by slightly turning the wide tube about its axis, when the rod is in vibration.

The vibrating rod and the air inside the tube are in stationary motion of the same frequency, the frequency being that of natural vibration of the rod, say n. The wave-length along the rod λ is approximately twice the length of the rod, that is, twice the distance between two antinodes; the wave-length in air λ_a is twice the distance between two nodes, as shown by the powder. Measure a distance including as many half wave-lengths as possible, and divide by the number of half wave-lengths.

But velocity of sound waves along rod $= V_r = n\lambda_r$, and velocity of sound waves through air in tube $= V_a = n\lambda_a$,

$$\therefore \quad \frac{V_r}{V_a} = \frac{n\lambda_r}{n\lambda_a} = \frac{\lambda_r}{\lambda_a}.$$

Knowing V_a and observing λ_a and λ_r, V_r may be calculated. Repeat for rods of brass and steel, using a resined cloth in place of the other for setting up the vibration.

(With longer rods it is preferable to clamp in two positions one quarter the length of the rod from each end as in the lower portion of Fig. 79. In such a case λ_r is equal to the length of the rod approximately.)

Young's Modulus.—Knowing the velocity of longitudinal vibrations along the rod, the value of Young's modulus (Y) of the material of the rod may be calculated from the formula $V_r = \sqrt{\frac{Y}{\rho}}$ where ρ is the density of the material of the rod. (Obtain ρ from tables or from experiments on a small portion of the material of the rod.)

Velocity of Sound in Gases.—If the end of the wider tube opposite to that enclosing the cardboard disc be fitted with a cork through which passes a glass tube, a supply of gas may be passed through the tube until the air has been almost entirely displaced. Repeat the above experiment with one of the rods used in the case of an air experiment, and again measure the distance between the nodes produced in the gas. In the air and gas experiments with the same vibrating rod, the frequency (n) is, of course, the same. The velocity of sound in the gas (V_g) is obtained from the equation $\frac{V_g}{V_a} = \frac{n\lambda_g}{n\lambda_a} = \frac{\lambda_g}{\lambda_a}$, by comparing the wave-lengths in the gas and in air.

VIII

TABLES.

USEFUL CONSTANTS

π	$= 3{\cdot}1416$	logarithm π	$= 0{\cdot}4972$
π^2	$= 9{\cdot}8696$	$\log \pi^2$	$= 0{\cdot}9943$
$1/\pi$	$= 0{\cdot}31831$	$\log 1/\pi$	$= \bar{1}{\cdot}5028$
$\sqrt{\pi}$	$= 1{\cdot}7725$	$\log \sqrt{\pi}$	$= 0{\cdot}2486$
$\log_{10}$	$= \log_e \times {\cdot}4343$	$\log_e$	$= \log_{10} \times 2{\cdot}3026$

1 inch	= 2·540 cm.	1 cm.	= 0·3937 inch
1 pound	= 453·6 grams	1 gram	= 15·432 grains
1 pint	= 0·5682 litre	1 litre	= 1·7598 pints
1 poundal	= 13825 dynes	1 dyne	= 0·00007233 poundal
1 foot-pound	= 1·356 joules	1 joule	= 0·7373 foot-pound
1 joule	= 10^7 ergs	1 calorie	= { 4·2 joules = $4{\cdot}2 \times 10^7$ ergs
1 horse power	= 746 watts	1 watt	= 0·00134 horse power
1 British thermal unit	= 252 calories	1 calorie	= 0·00397 B.T.U.

MOMENTS OF INERTIA

	Axis.	Moment of Inertia.
Uniform thin rod	through centre, perpendicular to axis of rod	$m \times \frac{l^2}{12}$
" rectangular plate	through centre, perp. to surface	$m \times \frac{l^2+b^2}{12}$
" rectangular plate	through centre, parallel to edge b	$m \times \frac{l^2}{12}$
" circular plate	through centre, perp. to surface	$m \times \frac{r^2}{2}$
" cylinder	axis of cylinder	$m \times \frac{r^2}{2}$
" circular plate	through centre, parallel to surface	$m \times \frac{r^2}{4}$
" sphere	through centre	$m \times \frac{2r^2}{5}$
" cylinder	through centre, perpendicular to axis	$m \times \left(\frac{l^2}{12}+\frac{r^2}{4}\right)$

M.I. about any axis parallel to and distant x from that in column 2, is found by adding mx^2 to value in column 3.

ACCELERATION OF GRAVITY (g.)

Place.	g.		
Pole	983·2	cm. per sec.^{2}	
Equator	978·1	,,	
Aberdeen	981·7	,,	
St Andrews	981·6	,,	
Edinburgh	981·5	,,	(32·2 ft. per sec.^{2})
Glasgow	981·6	,,	
London (Kew)	981·2	,,	

APPROXIMATE DENSITIES (in gr. per c.c.)

Aluminium	2·6	Brass	8·4
Copper	8·9	German silver	8·9
Gold	19·3	Quartz	2·65
Iron	7·8	Glass	2·5 to 3·5
Lead	11·4	Porcelain	2·3
Mercury	13·6	Ice	0·92
Nickel	8·9	Alcohol	0·81
Platinum	21·5	Ether	0·74
Silver	10·5	Glycerine	1·26
Tin	7·3	Paraffin oil	0·80
Zinc	7·1	Turpentine	0·87

DENSITY OF WATER AND MERCURY AT VARIOUS TEMPERATURES

Temp.	Water.	Mercury.
0	0·99987	13·5955
4	1·00000	13·5856
20	0·99823	13·5462
40	0·9922	13·4973
60	0·9832	13·4486
80	0·9718	13·4001
100	0·9584	13·3518

DENSITY OF DRY AIR AT DIFFERENT TEMPERATURES AND PRESSURES

Temperature.	73 cm. pressure.	74 cm. pressure.	75 cm. pressure.	76 cm. pressure.	77 cm. pressure.
10	·001198	·001214	·001231	·001247	·001264
12	·001190	·001206	·001222	·001238	·001255
14	·001181	·001197	·001214	·001230	·001246
16	·001173	·001189	·001205	·001221	·001237
18	·001165	·001181	·001197	·001213	·001229
20	·001157	·001173	·001189	·001205	·001220

ELASTICITY (C.G.S. units)

Substance.	Volume elasticity.	Rigidity.	Young's modulus.	Poisson's ratio.	Tenacity.
Copper . .	$1·3 \times 10^{12}$	$0·44 \times 10^{12}$	$1·24 \times 10^{12}$	·3	2×10^{9}
Iron (wrought)	$1·6 \times 10^{12}$	$0·8 \times 10^{12}$	$1·95 \times 10^{12}$	·28	3 to 4×10^{9}
Iron (cast) .	$0·95 \times 10^{12}$	$0·5 \times 10^{12}$	$1·3 \times 10^{12}$	·27	1 „ 2×10^{9}
Steel . .	$1·6 \times 10^{12}$	$0·8 \times 10^{12}$	$2·1 \times 10^{12}$	·28	2·5 „ 14×10^{9}
Brass . .	$1·1 \times 10^{12}$	$0·36 \times 10^{12}$	$1·0 \times 10^{12}$	·35	1·5 „ 4×10^{9}
Oak . .	...	...	$0·13 \times 10^{12}$	...	...
Deal . .	...	...	$0·09 \times 10^{12}$	...	...
Teak . .	...	...	$0·16 \times 10^{12}$	...	...

THERMAL CONSTANTS

Substance.	Coeff. of linear expansion.	Specific heat.	Thermal conductivity.
Aluminium . .	·000025	·21	0·50
Copper . .	·000017 to 19	·093	0·9
Iron . . .	·000012 „ 15	·11	0·16
Lead . . .	·000028	·031	0·08
Mercury . .	·00018 [cubic]	·033	0·016
Silver . . .	·000019	·056	1·0
Tin . . .	·000021	·056	0·15
Zinc . . .	·000027	·093	0·27
Brass . . .	·000018 to 19	·086	0·30
German silver .	·000018	·09	0·07
Ice . . .	·000051	·50	0·005
Paraffin wax .	·00012	·65	0·0005

Latent heat of fusion of ice = 80 calories per gram.
Latent heat of vaporisation of water at 100° C. = 537 calories per gram.
Latent heat of vaporisation of water at t° C., according to Regnault = 607 - ·7t
Mechanical equivalent of heat = $4·18 \times 10^7$ ergs per calorie.
= 776 ft.-pounds (Greenwich) per British unit of heat (water lb. deg. F.).

SURFACE TENSION AND ANGLE OF CONTACT WITH GLASS (approximately)

	Surface tension.	Contact angle.		Surface tension.	Contact angle.
	Dynes per cm.	Degrees.		Dynes per cm.	Degrees.
Water (15° C.) .	74	0	Benzene .	29	0
Mercury (15° C.) .	546	127	Ether (ethyl)	16	16
Alcohol (methyl) .	23	0	Paraffin oil .	26	26
Alcohol (ethyl) .	22	0	Turpentine .	27	17

SATURATED VAPOUR PRESSURE OF WATER

Temperature.	Pressure in cm. of mercury.	Temperature.	Pressure in cm. of mercury.
0	0·458	16	1·36
2	0·529	18	1·55
4	0·61	20	1·75
6	0·70	40	5·5
8	0·80	60	14·9
10	0·92	80	35·5
12	1·05	100	76·0
14	1·2		

GLAISHER'S FACTOR FOR WET AND DRY BULB HYGROMETER

$$f = \frac{\text{depression of dew point}}{\text{difference between wet and dry bulb readings}}.$$

Temperature of dry bulb	*f*.	Temperature of dry bulb.	*f*.
0	3·32	14	1·92
2	2·54	16	1·87
4	2·31	18	1·83
6	2·21	20	1·79
8	2·13	22	1·75
10	2·06	24	1·72
12	1·99		

REFRACTIVE INDEX (Sodium Light)

Substance	Index	Substance	Index
Light crown glass . .	1·51	Rock salt	1·54
Dense silicate flint glass .	1·65	Water	1·33

Velocity of light *in vacuo* = 3×10^{10} cm. per second.
= 186,400 miles per second.

ELECTROMOTIVE FORCES OF COMMON CELLS

Cell	Volts.	Cell	Volts.
Bichromate . .	about 2·0	Leclanché . .	about 1·5
Bunsen . . .	1·8	Secondary . .	2·2 to 1·9
Clark . . .	1·43	Weston . . .	1·02
Daniell . . .	1·1		

SPECIFIC RESISTANCE AND TEMPERATURE COEFFICIENT

Substance.	Specific resistance.	Temperature coefficient
Aluminium . .	3×10^{-6} c.g.s.	·0038
Copper . . .	1·8 "	·0043
Gold	2·4 "	·0040
Iron	12 to 14 "	·0062
Lead . . .	21 "	·0043
Mercury . . .	95 "	·0009
Silver . . .	1·6 "	·0040
Tin	11 "	·0045
Zinc	6·1 "	·0037
Brass . . .	6·8 "	·0010
German silver . .	16 to 40 "	·0002 to 6
Manganin . .	43 "	·000002 to 5
Platinoid . . .	34 "	·00025

TERRESTRIAL MAGNETIC CONSTANTS (1914)

Place.	Declination α	Inclination. δ.	Horizontal intensity. H.	Vertical intensity. V.
			c.g.s.	c.g.s.
Aberdeen .	16° 5′ W.	70° 35′	·163	·464
Eskdalemuir .	18° W.	69° 35′	·168	·452
Greenwich .	15° 20′ W.	66° 50′	·185	·434
Kew . .	15° 40′ W.	66° 55′	·185	·436

Logarithms.

	0	1	2	3	4	5	6	7	8	9	1	2	3	4	5	6	7	8	9
100	0000	0004	0009	0013	0017	0022	0026	0030	0035	0039									
101	0043	0048	0052	0056	0060	0065	0069	0073	0077	0082									
102	0086	0090	0095	0099	0103	0107	0111	0116	0120	0124									
103	0128	0133	0137	0141	0145	0149	0154	0158	0162	0166									
104	0170	0175	0179	0183	0187	0191	0195	0199	0204	0208									
105	0212	0216	0220	0224	0228	0233	0237	0241	0245	0249									
106	0253	0257	0261	0265	0269	0273	0278	0282	0286	0290									
107	0294	0298	0302	0306	0310	0314	0318	0322	0326	0330									
108	0334	0338	0342	0346	0350	0354	0358	0362	0366	0370									
109	0374	0378	0382	0386	0390	0394	0398	0402	0406	0410									
110	0414	0418	0422	0426	0430	0434	0438	0441	0445	0449									
11	0414	0453	0492	0531	0569	0607	0645	0682	0719	0755	4	8	11	15	19	23	26	30	34
12	0792	0828	0864	0899	0934	0969	1004	1038	1072	1106	3	7	10	14	17	21	24	28	31
13	1139	1173	1206	1239	1271	1303	1335	1367	1399	1430	3	6	10	13	16	19	23	26	29
14	1461	1492	1523	1553	1584	1614	1644	1673	1703	1732	3	6	9	12	15	18	21	24	27
15	1761	1790	1818	1847	1875	1903	1931	1959	1987	2014	3	6	8	11	14	17	20	22	25
16	2041	2068	2095	2122	2148	2175	2201	2227	2253	2279	3	5	8	11	13	16	18	21	24
17	2304	2330	2355	2380	2405	2430	2455	2480	2504	2529	2	5	7	10	12	15	17	20	22
18	2553	2577	2601	2625	2648	2672	2695	2718	2742	2765	2	5	7	9	12	14	16	19	21
19	2788	2810	2833	2856	2878	2900	2923	2945	2967	2989	2	4	7	9	11	13	16	18	20
20	3010	3032	3054	3075	3096	3118	3139	3160	3181	3201	2	4	6	8	11	13	15	17	19
21	3222	3243	3263	3284	3304	3324	3345	3365	3385	3404	2	4	6	8	10	12	14	16	18
22	3424	3444	3464	3483	3502	3522	3541	3560	3579	3598	2	4	6	8	10	12	14	15	17
23	3617	3636	3655	3674	3692	3711	3729	3747	3766	3784	2	4	6	7	9	11	13	15	17
24	3802	3820	3838	3856	3874	3892	3909	3927	3945	3962	2	4	5	7	9	11	12	14	16
25	3979	3997	4014	4031	4048	4065	4082	4099	4116	4133	2	3	5	7	9	10	12	14	15
26	4150	4166	4183	4200	4216	4232	4249	4265	4281	4298	2	3	5	7	8	10	11	13	15
27	4314	4330	4346	4362	4378	4393	4409	4425	4440	4456	2	3	5	6	8	9	11	13	14
28	4472	4487	4502	4518	4533	4548	4564	4579	4594	4609	2	3	5	6	8	9	11	12	14
29	4624	4639	4654	4669	4683	4698	4713	4728	4742	4757	1	3	4	6	7	9	10	12	13
30	4771	4786	4800	4814	4829	4843	4857	4871	4886	4900	1	3	4	6	7	9	10	11	13
31	4914	4928	4942	4955	4969	4983	4997	5011	5024	5038	1	3	4	6	7	8	10	11	12
32	5051	5065	5079	5092	5105	5119	5132	5145	5159	5172	1	3	4	5	7	8	9	11	12
33	5185	5198	5211	5224	5237	5250	5263	5276	5289	5302	1	3	4	5	6	8	9	10	12
34	5315	5328	5340	5353	5366	5378	5391	5403	5416	5428	1	3	4	5	6	8	9	10	11
35	5441	5453	5465	5478	5490	5502	5514	5527	5539	5551	1	2	4	5	6	7	9	10	11
36	5563	5575	5587	5599	5611	5623	5635	5647	5658	5670	1	2	4	5	6	7	8	10	11
37	5682	5694	5705	5717	5729	5740	5752	5763	5775	5786	1	2	3	5	6	7	8	9	10
38	5798	5809	5821	5832	5843	5855	5866	5877	5888	5899	1	2	3	5	6	7	8	9	10
39	5911	5922	5933	5944	5955	5966	5977	5988	5999	6010	1	2	3	4	5	7	8	9	10
40	6021	6031	6042	6053	6064	6075	6085	6096	6107	6117	1	2	3	4	5	6	8	9	10
41	6128	6138	6149	6160	6170	6180	6191	6201	6212	6222	1	2	3	4	5	6	7	8	9
42	6232	6243	6253	6263	6274	6284	6294	6304	6314	6325	1	2	3	4	5	6	7	8	9
43	6335	6345	6355	6365	6375	6385	6395	6405	6415	6425	1	2	3	4	5	6	7	8	9
44	6435	6444	6454	6464	6474	6484	6493	6503	6513	6522	1	2	3	4	5	6	7	8	9
45	6532	6542	6551	6561	6571	6580	6590	6599	6609	6618	1	2	3	4	5	6	7	8	9
46	6628	6637	6646	6656	6665	6675	6684	6693	6702	6712	1	2	3	4	5	6	7	7	8
47	6721	6730	6739	6749	6758	6767	6776	6785	6794	6803	1	2	3	4	5	5	6	7	8
48	6812	6821	6830	6839	6848	6857	6866	6875	6884	6893	1	2	3	4	4	5	6	7	8
49	6902	6911	6920	6928	6937	6946	6955	6964	6972	6981	1	2	3	4	4	5	6	7	8

Logarithms.

	0	1	2	3	4	5	6	7	8	9	1	2	3	4	5	6	7	8	9
50	6990	6998	7007	7016	7024	7033	7042	7050	7059	7067	1	2	3	3	4	5	6	7	8
51	7076	7084	7093	7101	7110	7118	7126	7135	7143	7152	1	2	3	3	4	5	6	7	8
52	7160	7168	7177	7185	7193	7202	7210	7218	7226	7235	1	2	2	3	4	5	6	7	7
53	7243	7251	7259	7267	7275	7284	7292	7300	7308	7316	1	2	2	3	4	5	6	6	7
54	7324	7332	7340	7348	7356	7364	7372	7380	7388	7396	1	2	2	3	4	5	6	6	7
55	7404	7412	7419	7427	7435	7443	7451	7459	7466	7474	1	2	2	3	4	5	5	6	7
56	7482	7490	7497	7505	7513	7520	7528	7536	7543	7551	1	2	2	3	4	5	5	6	7
57	7559	7566	7574	7582	7589	7597	7604	7612	7619	7627	1	2	2	3	4	5	5	6	7
58	7634	7642	7649	7657	7664	7672	7679	7686	7694	7701	1	1	2	3	4	4	5	6	7
59	7709	7716	7723	7731	7738	7745	7752	7760	7767	7774	1	1	2	3	4	4	5	6	7
60	7782	7789	7796	7803	7810	7818	7825	7832	7839	7846	1	1	2	3	4	4	5	6	6
61	7853	7860	7868	7875	7882	7889	7896	7903	7910	7917	1	1	2	3	4	4	5	6	6
62	7924	7931	7938	7945	7952	7959	7966	7973	7980	7987	1	1	2	3	3	4	5	6	6
63	7993	8000	8007	8014	8021	8028	8035	8041	8048	8055	1	1	2	3	3	4	5	5	6
64	8062	8069	8075	8082	8089	8096	8102	8109	8116	8122	1	1	2	3	3	4	5	5	6
65	8129	8136	8142	8149	8156	8162	8169	8176	8182	8189	1	1	2	3	3	4	5	5	6
66	8195	8202	8209	8215	8222	8228	8235	8241	8248	8254	1	1	2	3	3	4	5	5	6
67	8261	8267	8274	8280	8287	8293	8299	8306	8312	8319	1	1	2	3	3	4	5	5	6
68	8325	8331	8338	8344	8351	8357	8363	8370	8376	8382	1	1	2	3	3	4	4	5	6
69	8388	8395	8401	8407	8414	8420	8426	8432	8439	8445	1	1	2	2	3	4	4	5	6
70	8451	8457	8463	8470	8476	8482	8488	8494	8500	8506	1	1	2	2	3	4	4	5	6
71	8513	8519	8525	8531	8537	8543	8549	8555	8561	8567	1	1	2	2	3	4	4	5	5
72	8573	8579	8585	8591	8597	8603	8609	8615	8621	8627	1	1	2	2	3	4	4	5	5
73	8633	8639	8645	8651	8657	8663	8669	8675	8681	8686	1	1	2	2	3	4	4	5	5
74	8692	8698	8704	8710	8716	8722	8727	8733	8739	8745	1	1	2	2	3	4	4	5	5
75	8751	8756	8762	8768	8774	8779	8785	8791	8797	8802	1	1	2	2	3	3	4	5	5
76	8808	8814	8820	8825	8831	8837	8842	8848	8854	8859	1	1	2	2	3	3	4	5	5
77	8865	8871	8876	8882	8887	8893	8899	8904	8910	8915	1	1	2	2	3	3	4	4	5
78	8921	8927	8932	8938	8943	8949	8954	8960	8965	8971	1	1	2	2	3	3	4	4	5
79	8976	8982	8987	8993	8998	9004	9009	9015	9020	9025	1	1	2	2	3	3	4	4	5
80	9031	9036	9042	9047	9053	9058	9063	9069	9074	9079	1	1	2	2	3	3	4	4	5
81	9085	9090	9096	9101	9106	9112	9117	9122	9128	9133	1	1	2	2	3	3	4	4	5
82	9138	9143	9149	9154	9159	9165	9170	9175	9180	9186	1	1	2	2	3	3	4	4	5
83	9191	9196	9201	9206	9212	9217	9222	9227	9232	9238	1	1	2	2	3	3	4	4	5
84	9243	9248	9253	9258	9263	9269	9274	9279	9284	9289	1	1	2	2	3	3	4	4	5
85	9294	9299	9304	9309	9315	9320	9325	9330	9335	9340	1	1	2	2	3	3	4	4	5
86	9345	9350	9355	9360	9365	9370	9375	9380	9385	9390	1	1	2	2	3	3	4	4	5
87	9395	9400	9405	9410	9415	9420	9425	9430	9435	9440	0	1	1	2	2	3	3	4	4
88	9445	9450	9455	9460	9465	9469	9474	9479	9484	9489	0	1	1	2	2	3	3	4	4
89	9494	9499	9504	9509	9513	9518	9523	9528	9533	9538	0	1	1	2	2	3	3	4	4
90	9542	9547	9552	9557	9562	9566	9571	9576	9581	9586	0	1	1	2	2	3	3	4	4
91	9590	9595	9600	9605	9609	9614	9619	9624	9628	9633	0	1	1	2	2	3	3	4	4
92	9638	9643	9647	9652	9657	9661	9666	9671	9675	9680	0	1	1	2	2	3	3	4	4
93	9685	9689	9694	9699	9703	9708	9713	9717	9722	9727	0	1	1	2	2	3	3	4	4
94	9731	9736	9741	9745	9750	9754	9759	9763	9768	9773	0	1	1	2	2	3	3	4	4
95	9777	9782	9786	9791	9795	9800	9805	9809	9814	9818	0	1	1	2	2	3	3	4	4
96	9823	9827	9832	9836	9841	9845	9850	9854	9859	9863	0	1	1	2	2	3	3	4	4
97	9868	9872	9877	9881	9886	9890	9894	9899	9903	9908	0	1	1	2	2	3	3	4	4
98	9912	9917	9921	9926	9930	9934	9939	9943	9948	9952	0	1	1	2	2	3	3	4	4
99	9956	9961	9965	9969	9974	9978	9983	9987	9991	9996	0	1	1	2	2	3	3	3	4

ANTILOGARITHMS

	0	1	2	3	4	5	6	7	8	9	Mean Differences.								
											1	2	3	4	5	6	7	8	9
·00	1000	1002	1005	1007	1009	1012	1014	1016	1019	1021	0	0	1	1	1	1	2	2	2
·01	1023	1026	1028	1030	1033	1035	1038	1040	1042	1045	0	0	1	1	1	1	2	2	2
·02	1047	1050	1052	1054	1057	1059	1062	1064	1067	1069	0	0	1	1	1	1	2	2	2
·03	1072	1074	1076	1079	1081	1084	1086	1089	1091	1094	0	0	1	1	1	1	2	2	2
·04	1096	1099	1102	1104	1107	1109	1112	1114	1117	1119	0	1	1	1	1	2	2	2	2
·05	1122	1125	1127	1130	1132	1135	1138	1140	1143	1146	0	1	1	1	1	2	2	2	2
·06	1148	1151	1153	1156	1159	1161	1164	1167	1169	1172	0	1	1	1	1	2	2	2	2
·07	1175	1178	1180	1183	1186	1189	1191	1194	1197	1199	0	1	1	1	1	2	2	2	2
·08	1202	1205	1208	1211	1213	1216	1219	1222	1225	1227	0	1	1	1	1	2	2	2	3
·09	1230	1233	1236	1239	1242	1245	1247	1250	1253	1256	0	1	1	1	1	2	2	2	3
·10	1259	1262	1265	1268	1271	1274	1276	1279	1282	1285	0	1	1	1	1	2	2	2	3
·11	1288	1291	1294	1297	1300	1303	1306	1309	1312	1315	0	1	1	1	2	2	2	2	3
·12	1318	1321	1324	1327	1330	1334	1337	1340	1343	1346	0	1	1	1	2	2	2	2	3
·13	1349	1352	1355	1358	1361	1365	1368	1371	1374	1377	0	1	1	1	2	2	2	3	3
·14	1380	1384	1387	1390	1393	1396	1400	1403	1406	1409	0	1	1	1	2	2	2	3	3
·15	1413	1416	1419	1422	1426	1429	1432	1435	1439	1442	0	1	1	1	2	2	2	3	3
·16	1445	1449	1452	1455	1459	1462	1466	1469	1472	1476	0	1	1	1	2	2	2	3	3
·17	1479	1483	1486	1489	1493	1496	1500	1503	1507	1510	0	1	1	1	2	2	2	3	3
·18	1514	1517	1521	1524	1528	1531	1535	1538	1542	1545	0	1	1	1	2	2	2	3	3
·19	1549	1552	1556	1560	1563	1567	1570	1574	1578	1581	0	1	1	1	2	2	3	3	3
·20	1585	1589	1592	1596	1600	1603	1607	1611	1614	1618	0	1	1	1	2	2	3	3	3
·21	1622	1626	1629	1633	1637	1641	1644	1648	1652	1656	0	1	1	2	2	2	3	3	3
·22	1660	1663	1667	1671	1675	1679	1683	1687	1690	1694	0	1	1	2	2	2	3	3	3
·23	1698	1702	1706	1710	1714	1718	1722	1726	1730	1734	0	1	1	2	2	2	3	3	4
·24	1738	1742	1746	1750	1754	1758	1762	1766	1770	1774	0	1	1	2	2	2	3	3	4
·25	1778	1782	1786	1791	1795	1799	1803	1807	1811	1816	0	1	1	2	2	2	3	3	4
·26	1820	1824	1828	1832	1837	1841	1845	1849	1854	1858	0	1	1	2	2	3	3	3	4
·27	1862	1866	1871	1875	1879	1884	1888	1892	1897	1901	0	1	1	2	2	3	3	3	4
·28	1905	1910	1914	1919	1923	1928	1932	1936	1941	1945	0	1	1	2	2	3	3	4	4
·29	1950	1954	1959	1963	1968	1972	1977	1982	1986	1991	0	1	1	2	2	3	3	4	4
·30	1995	2000	2004	2009	2014	2018	2023	2028	2032	2037	0	1	1	2	2	3	3	4	4
·31	2042	2046	2051	2056	2061	2065	2070	2075	2080	2084	0	1	1	2	2	3	3	4	4
·32	2089	2094	2099	2104	2109	2113	2118	2123	2128	2133	0	1	1	2	2	3	3	4	4
·33	2138	2143	2148	2153	2158	2163	2168	2173	2178	2183	0	1	1	2	2	3	3	4	4
·34	2188	2193	2198	2203	2208	2213	2218	2223	2228	2234	1	1	2	2	3	3	4	4	5
·35	2239	2244	2249	2254	2259	2265	2270	2275	2280	2286	1	1	2	2	3	3	4	4	5
·36	2291	2296	2301	2307	2312	2317	2323	2328	2333	2339	1	1	2	2	3	3	4	4	5
·37	2344	2350	2355	2360	2366	2371	2377	2382	2388	2393	1	1	2	2	3	3	4	4	5
·38	2399	2404	2410	2415	2421	2427	2432	2438	2443	2449	1	1	2	2	3	3	4	4	5
·39	2455	2460	2466	2472	2477	2483	2489	2495	2500	2506	1	1	2	2	3	3	4	5	5
·40	2512	2518	2523	2529	2535	2541	2547	2553	2559	2564	1	1	2	2	3	4	4	5	5
·41	2570	2576	2582	2588	2594	2600	2606	2612	2618	2624	1	1	2	2	3	4	4	5	5
·42	2630	2636	2642	2649	2655	2661	2667	2673	2679	2685	1	1	2	2	3	4	4	5	6
·43	2692	2698	2704	2710	2716	2723	2729	2735	2742	2748	1	1	2	3	3	4	4	5	6
·44	2754	2761	2767	2773	2780	2786	2793	2799	2805	2812	1	1	2	3	3	4	4	5	6
·45	2818	2825	2831	2838	2844	2851	2858	2864	2871	2877	1	1	2	3	3	4	5	5	6
·46	2884	2891	2897	2904	2911	2917	2924	2931	2938	2944	1	1	2	3	3	4	5	5	6
·47	2951	2958	2965	2972	2979	2985	2992	2999	3006	3013	1	1	2	3	3	4	5	5	6
·48	3020	3027	3034	3041	3048	3055	3062	3069	3076	3083	1	1	2	3	4	4	5	6	6
·49	3090	3097	3105	3112	3119	3126	3133	3141	3148	3155	1	1	2	3	4	4	5	6	6

ANTILOGARITHMS

	0	1	2	3	4	5	6	7	8	9	Mean Differences.								
											1	2	3	4	5	6	7	8	9
·50	3162	3170	3177	3184	3192	3199	3206	3214	3221	3228	1	1	2	3	4	4	5	6	7
·51	3236	3243	3251	3258	3266	3273	3281	3289	3296	3304	1	2	2	3	4	5	5	6	7
·52	3311	3319	3327	3334	3342	3350	3357	3365	3373	3381	1	2	2	3	4	5	5	6	7
·53	3388	3396	3404	3412	3420	3428	3436	3443	3451	3459	1	2	2	3	4	5	6	6	7
·54	3467	3475	3483	3491	3499	3508	3516	3524	3532	3540	1	2	2	3	4	5	6	6	7
·55	3548	3556	3565	3573	3581	3589	3597	3606	3614	3622	1	2	2	3	4	5	6	7	7
·56	3631	3639	3648	3656	3664	3673	3681	3690	3698	3707	1	2	3	3	4	5	6	7	8
·57	3715	3724	3733	3741	3750	3758	3767	3776	3784	3793	1	2	3	3	4	5	6	7	8
·58	3802	3811	3819	3828	3837	3846	3855	3864	3873	3882	1	2	3	4	4	5	6	7	8
·59	3890	3899	3908	3917	3926	3936	3945	3954	3963	3972	1	2	3	4	5	5	6	7	8
·60	3981	3990	3999	4009	4018	4027	4036	4046	4055	4064	1	2	3	4	5	6	6	7	8
·61	4074	4083	4093	4102	4111	4121	4130	4140	4150	4159	1	2	3	4	5	6	7	8	9
·62	4169	4178	4188	4198	4207	4217	4227	4236	4246	4256	1	2	3	4	5	6	7	8	9
·63	4266	4276	4285	4295	4305	4315	4325	4335	4345	4355	1	2	3	4	5	6	7	8	9
·64	4365	4375	4385	4395	4406	4416	4426	4436	4446	4457	1	2	3	4	5	6	7	8	9
·65	4467	4477	4487	4498	4508	4519	4529	4539	4550	4560	1	2	3	4	5	6	7	8	9
·66	4571	4581	4592	4603	4613	4624	4634	4645	4656	4667	1	2	3	4	5	6	7	9	10
·67	4677	4688	4699	4710	4721	4732	4742	4753	4764	4775	1	2	3	4	5	7	8	9	10
·68	4786	4797	4808	4819	4831	4842	4853	4864	4875	4887	1	2	3	4	6	7	8	9	10
·69	4898	4909	4920	4932	4943	4955	4966	4977	4989	5000	1	2	3	5	6	7	8	9	10
·70	5012	5023	5035	5047	5058	5070	5082	5093	5105	5117	1	2	4	5	6	7	8	9	11
·71	5129	5140	5152	5164	5176	5188	5200	5212	5224	5236	1	2	4	5	6	7	8	10	11
·72	5248	5260	5272	5284	5297	5309	5321	5333	5346	5358	1	2	4	5	6	7	9	10	11
·73	5370	5383	5395	5408	5420	5433	5445	5458	5470	5483	1	3	4	5	6	8	9	10	11
·74	5495	5508	5521	5534	5546	5559	5572	5585	5598	5610	1	3	4	5	6	8	9	10	12
·75	5623	5636	5649	5662	5675	5689	5702	5715	5728	5741	1	3	4	5	7	8	9	10	12
·76	5754	5768	5781	5794	5808	5821	5834	5848	5861	5875	1	3	4	5	7	8	9	11	12
·77	5888	5902	5916	5929	5943	5957	5970	5984	5998	6012	1	3	4	5	7	8	10	11	12
·78	6026	6039	6053	6067	6081	6095	6109	6124	6138	6152	1	3	4	6	7	8	10	11	13
·79	6166	6180	6194	6209	6223	6237	6252	6266	6281	6295	1	3	4	6	7	9	10	11	13
·80	6310	6324	6339	6353	6368	6383	6397	6412	6427	6442	1	3	4	6	7	9	10	12	13
·81	6457	6471	6486	6501	6516	6531	6546	6561	6577	6592	2	3	5	6	8	9	11	12	14
·82	6607	6622	6637	6653	6668	6683	6699	6714	6730	6745	2	3	5	6	8	9	11	12	14
·83	6761	6776	6792	6808	6823	6839	6855	6871	6887	6902	2	3	5	6	8	9	11	13	14
·84	6918	6934	6950	6966	6982	6998	7015	7031	7047	7063	2	3	5	6	8	10	11	13	15
·85	7079	7096	7112	7129	7145	7161	7178	7194	7211	7228	2	3	5	7	8	10	12	13	15
·86	7244	7261	7278	7295	7311	7328	7345	7362	7379	7396	2	3	5	7	8	10	12	13	15
·87	7413	7430	7447	7464	7482	7499	7516	7534	7551	7568	2	3	5	7	9	10	12	14	16
·88	7586	7603	7621	7638	7656	7674	7691	7709	7727	7745	2	4	5	7	9	11	12	14	16
·89	7762	7780	7798	7816	7834	7852	7870	7889	7907	7925	2	4	5	7	9	11	13	14	16
·90	7943	7962	7980	7998	8017	8035	8054	8072	8091	8110	2	4	6	7	9	11	13	15	17
·91	8128	8147	8166	8185	8204	8222	8241	8260	8279	8299	2	4	6	8	9	11	13	15	17
·92	8318	8337	8356	8375	8395	8414	8433	8453	8472	8492	2	4	6	8	10	12	14	15	17
·93	8511	8531	8551	8570	8590	8610	8630	8650	8670	8690	2	4	6	8	10	12	14	16	18
·94	8710	8730	8750	8770	8790	8810	8831	8851	8872	8892	2	4	6	8	10	12	14	16	18
·95	8913	8933	8954	8974	8995	9016	9036	9057	9078	9099	2	4	6	8	10	12	15	17	19
·96	9120	9141	9162	9183	9204	9226	9247	9268	9290	9311	2	4	6	8	11	13	15	17	19
·97	9333	9354	9376	9397	9419	9441	9462	9484	9506	9528	2	4	7	9	11	13	15	17	20
·98	9550	9572	9594	9616	9638	9661	9683	9705	9727	9750	2	4	7	9	11	13	16	18	20
·99	9772	9795	9817	9840	9863	9886	9908	9931	9954	9977	2	5	7	9	11	14	16	18	20

TABLE OF TRIGONOMETRICAL RATIOS

	Sine.	Cosine.	Tangent.	Radians.		Sine.	Cosine.	Tangent.	Radians.
0°	·0000	1·000	·0000	·0000	45°	·7071	·7071	1·0000	·7854
1	·0175	·9998	·0175	·0175	46	·7193	·6947	1·0355	·8029
2	·0349	·9994	·0349	·0349	47	·7314	·6820	1·0724	·8203
3	·0523	·9986	·0524	·0524	48	·7431	·6691	1·1106	·8378
4	·0698	·9976	·0699	·0698	49	·7547	·6561	1·1504	·8552
5	·0872	·9962	·0875	·0873	50°	·7660	·6428	1·1918	·8727
6	·1045	·9945	·1051	·1047	51	·7771	·6293	1·2349	·8901
7	·1219	·9925	·1228	·1222	52	·7880	·6157	1·2799	·9076
8	·1392	·9903	·1405	·1396	53	·7986	·6018	1·3270	·9250
9	·1564	·9877	·1584	·1571	54	·8090	·5878	1·3764	·9425
10°	·1736	·9848	·1763	·1745	55	·8192	·5736	1·4281	·9599
11	·1908	·9816	·1944	·1920	56	·8290	·5592	1·4826	·9774
12	·2079	·9781	·2126	·2094	57	·8387	·5446	1·5399	·9948
13	·2250	·9744	·2309	·2269	58	·8480	·5299	1·6003	1·0123
14	·2419	·9703	·2493	·2443	59	·8572	·5150	1·6643	1·0297
15	·2588	·9659	·2679	·2618	60°	·8660	·5000	1·7321	1·0472
16	·2756	·9613	·2867	·2793	61	·8746	·4848	1·8040	1·0647
17	·2924	·9563	·3057	·2967	62	·8829	·4695	1·8807	1·0821
18	·3090	·9511	·3249	·3142	63	·8910	·4540	1·9626	1·0996
19	·3256	9455	·3443	·3316	64	·8988	·4384	2·0503	1·1170
20°	·3420	·9397	·3640	·3491	65	·9063	·4226	2·1445	1·1345
21	·3584	·9336	·3839	·3665	66	·9135	·4067	2·2460	1·1519
22	·3746	·9272	·4040	·3840	67	·9205	·3907	2·3559	1·1694
23	·3907	·9205	·4245	·4014	68	·9272	·3746	2·4751	1·1868
24	·4067	·9135	·4452	·4189	69	·9336	·3584	2·6051	1·2043
25	·4226	·9063	·4663	·4363	70°	·9397	·3420	2·7475	1·2217
26	·4384	·8988	·4877	·4538	71	·9455	·3256	2·9042	1·2392
27	·4540	·8910	·5095	·4712	72	·9511	·3090	3·0777	1·2566
28	·4695	·8829	·5317	·4887	73	·9563	·2924	3·2709	1·2741
29	·4848	·8746	·5543	·5061	74	·9613	·2756	3·4874	1·2915
30°	·5000	·8660	·5774	·5236	75	·9659	·2588	3·7321	1·3090
31	·5150	·8572	·6009	·5411	76	·9703	·2419	4 0108	1·3265
32	·5299	·8480	·6249	·5585	77	·9744	·2250	4·3315	1·3439
33	·5446	·8387	·6494	·5760	78	·9781	·2079	4·7046	1·3614
34	·5592	·8290	·6745	·5934	79	·9816	·1908	5·1446	1·3788
35	·5736	·8192	·7002	·6109	80°	·9848	·1736	5·6713	1·3963
36	·5878	·8090	·7265	·6283	81	·9877	·1564	6·3138	1·4137
37	·6018	·7986	·7536	·6458	82	·9903	·1392	7·1154	1·4312
38	·6157	·7880	·7813	·6632	83	·9925	·1219	8·1443	1·4486
39	·6293	·7771	·8098	·6807	84	·9945	·1045	9·514	1·4661
40°	·6428	·7660	·8391	·6981	85	·9962	·0872	11·43	1·4835
41	·6561	·7547	·8693	·7156	86	·9976	·0698	14·30	1·5010
42	·6691	·7431	·9004	·7330	87	·9986	·0523	19·08	1·5184
43	·6820	·7314	·9325	·7505	88	·9994	·0349	28·64	1·5359
44	·6947	·7193	·9657	·7679	89	·9998	·0175	57·29	1·5533